Promises of the Political

Promises of the Political

Insurgent Cities in a Post-Political Environment

Erik Swyngedouw

The MIT Press
Cambridge, Massachusetts
London, England

This book was set in ITC Stone by Jen Jackowitz.

Library of Congress Cataloging-in-Publication Data is available.

ISBN: 978-0-262-03822-5 (hc); 978-0-262-53565-6 (pb)

To all insurgent citizens who remain faithful to the necessity of an equal, free, and solidarity-based urbanity in an ecologically sane world

The people is those who, refusing to be the population, disrupt the system.

—Michel Foucault

Contents

List of Acronyms

ANC	African National Congress
BP	British Petroleum
CETA	Comprehensive Economic Trade Agreement (between the EU and Canada)
COP	Conference of the Parties (of the United Nations Framework Convention on Climate Change)
DNA	Deoxyribonucleic acid
EU	European Union
GMO	Genetically modified organism
HIV	Human immunodeficiency virus
IMF	International Monetary Fund
IPCC	Intergovernmental Panel on Climate Change
NASA	National Aeronautics and Space Administration
NIMBY	Not In My Back Yard
NGO	Non-governmental organizations
TTIP	Transatlantic Trade and Investment Partnership (between the EU and the US)
UEJ	Urban environmental justice
UPE	Urban political ecology
UK	United Kingdom
US	United States
WTO	World Trade Organization

Preface

The manuscript of this book was completed during a time of major upheaval. A series of terrorist attacks sent shockwaves through the world's cities and disrupted the choreography of urban life in many parts of the world. Trump became president of the United States and announced his withdrawal from the Paris Climate Agreement alongside a range of other rather disturbing geopolitical initiatives. In post-Brexit UK, a general sentiment of doom and gloom prevails among many who fear Britain's descent into a national populist isolationism. Turkey, Poland, Hungary, and a range of other countries decidedly march in the direction of overtly authoritarian political rule. Xenophobic and right wing parties make great strides. In Germany, the "Alternative for Germany," a radical nationalist party, entered parliament. Such disturbing geopolitical trends have not been seen since the Second World War. Tales of corruption and the abuse of power by both economic and political elites make daily media headlines. Eight years after the deepest economic crisis since at least the 1930s threw millions of people into uncertainty, precarity, and poverty, the economic elites are doing very well as the frenetic dance of a globalized financial-economic system has resumed its feral drive for earthly gain. In the meantime, most ecological parameters continue to worsen as the world's ecologies undergo rapid and probably irreversible change. When I started this book a few years ago, I could not have imagined that things could have become worse than they were then. But of course, as Hamlet already insisted, "There are more things in heaven and earth, Horatio, than are dreamt of in your philosophy."

Nonetheless, I encounter in the streets and squares of our cities and neighborhoods extraordinary people who mobilize their often-formidable capacities to fight for a different, ecologically saner, and socioeconomically more equal world. The many social experiments and political movements that have been dotting the urban landscape over the past few years are indeed testimony to the desire of many to mobilize actively for a political transformation that seeks to construct more livable cities in a more equal world, where people can pursue their lives in free association with others. This book is dedicated to this motley crew of activists who in a variety of ways share the necessity to fight for a different and more inclusive world.

The book took several years to complete. Several chapters have been published in some form or other before, but they have been updated and thoroughly rewritten to fit the purposes of this book with permission obtained from all the relevant publishers.[1] I am very grateful to the editors and publishers who granted permission to use (part of) the material published here. In the long process of putting this book together, I have been indebted to a large number of people who were instrumental in both helping to shape the ideas developed here and keeping me sane when there was a danger of things spiraling out of control. I am really grateful to each and every one of them. In particular, my partner Maria Kaika has been a continuous source of support, critical engagement, and encouragement. The great bunch of ENTITLE fellows provided a supportive and energizing network of like-minded scholar-activists. My graduate students and postdoctoral fellows at Manchester University, and a few elsewhere, were a continuous source of inspiration, while the university offered a space that nurtured intellectual engagement and provided a stimulating environment, both academically and socially. The research networks that Frank Moulaert sustained over many years offered a wonderful social home and an intellectual environment that fermented thinking and acting in politically and socially innovative manners. Eva, Nikolaas, and Arno accompanied me on our collective journey of trying, but failing, to really grow up. And of course, the millions of urban insurgents who put their bodies on the line in the fight for a more egalitarian and inclusive city and world were a continuous source of hope and inspiration.

Introduction: Promises of the Political

Western democracies are only the political facade of economic power. A facade with colours, banners, and endless debates about sacrosanct democracy. We live in an era where we can discuss everything. With one exception: Democracy. She is there, an acquired dogma. Don't touch, like a museum display. Elections have become the representation of an absurd comedy, shameful, where the participation of the citizen is very weak, and in which the governments represent the political commissionaires of economic power.
—José Saramago

I have never voted. Like most people I am utterly disenchanted by politics. Like most people I regard politicians as frauds and liars and the current political system as nothing more than a bureaucratic means for furthering the augmentation and advantages of economic elites
—Russell Brand

Over the past few years or so, the concept of the post-political has been developed and contested by a range of activists and scholars working across the social sciences, many of whom have sought to demonstrate its performativity as a critical tool for conceptualizing contemporary processes of depoliticization and repoliticization. Post-politicization refers to the contested and uneven process by which consensual governance of contentious public affairs through the mobilization of techno-managerial dispositives sutures or colonizes the space of the political. Such depoliticizing gestures disavow the inherently heterogeneous and often antagonistic relations that cut through the social, and reduce the terrain of the political to the art or *techné* of public management. In other words, the political domain has been systematically narrowed over the past few decades to a techno-managerial apparatus of governance whereby fundamental choices are no longer possible or deemed reasonable. While problems and contentious issues of public concern (like environmental crises, urban revolts, terrorist threats, or economic conditions) are generally recognized, they are dealt with by means of consensual governance arrangements that do not question the wider social, ecological, and political-economic frame. Technological, institutional, and managerial "fixes" are negotiated that leave the basic political-economic structure intact. Growing apathy from an increasingly disenfranchised public has paralleled this depoliticization, and resulted in a growing electoral appeal of antiestablishment and populist forces and political parties. Both political participation and trust in politics-as-usual is at an all-time low, particularly in the Global North. Such "retreat of the political" and the related schism between public governance and political participation are often noted, yet rarely analyzed. In addition, a wide range of urban protests and rebellions, sparked off by deep-seated discontent with current socioeconomic and ecological conditions, have sprung up in many cities around the world, from the Arab Spring, to the Spanish *Indignados* or the Occupy movement. These and other outbursts of political discontent suggest both the extent of boiling rancor and points at possible new forms of politicization.

The key question this book attempts to answer is if and how progressive and emancipatory politics can still be thought and practiced in the twenty-first century. Are we condemned to mobilize our humanitarian capacities to assuage the rougher edges of a parliamentary-capitalist order that reproduces social inequality, political exclusion, and ecological disintegration, or is it possible to think of a different socio-ecological organization articulated around equality and solidarity, and of new political practices that can spark off and sustain such political-ecological transformations? Ultimately, this is the challenge of our time. Those who still maintain fidelity to an emancipatory project can subscribe to either the consensualizing logic of a post-democratic constellation or organize around the belief that a more equal socio-ecological ordering is both practical and necessary. It is within this tension that sides have to be taken. The task of this book is to excavate the conditions under which these choices have to be made and the possibilities lurking within the interstices of the present order.

Part I of this book explores the contours of post-politicization, and proposes a series of theoretical approaches to excavating the dynamics through which post-politicizing modes of governing have come into being over the past few decades. Through the emblematic lens of a critical engagement with both the urban and environmental condition we are in, part II charts the political-ecological and socioeconomic processes through which the question of the political became evacuated from public space and integrated within post-democratic institutional arrangements. In part III, the book identifies a range of new forms of politicization that mark the present geopolitical landscape in ways that potentially open up an incipient "return of the political" under the universalizing banner of freedom, equality, and solidarity, ushering in new forms of being-in-common in democratizing manners. I shall argue that the incipient return of the political as manifested in the proliferating urban revolts and uprisings that have choreographed urban political dynamics in many places in the world since 2011 may signal a revival of the emancipatory and democratizing impulse.

Indeed, we live in times both haunted and paradoxical. Instituted representational democracy is more widespread than ever. Identitarian concerns related to religious beliefs, race, ethnicity, sexuality, and other preferences, and all manner of issues and problems are made visible and politicized. "Participatory" and "inclusive" forms of "networked" governance are nurtured and fostered at a range of geographical scales. Lifestyle choices, the unsustainable reengineering of our climate, the sexual escapades of the former IMF chairman, the heroic resistances of indigenous peoples against dispossession and resource extraction, the repression of gay people in Russia, the garbage left on my sidewalk, the plight of the whale, the governments' austerity agendas to get the economy out of the doldrums—all these issues and an infinity of others are seemingly politicized. That is, they are discussed, researched, dissected, evaluated, raised to issues of public concern, and debated at length in a variety of public and political arenas. Everything, so it seems, can be aired, made visible, and rendered contentious.

In short, democracy as the theater of and for the pluralistic and disputed consideration of matters of public concern would appear to be triumphant. Political elites, irrespective of their particular party allegiance, do not tire of pointing out the great strides that democratic civic life has made over the past few decades. We are told that the great battle of the twentieth century between totalitarianism and democracy has been finally and decisively concluded. The history of humanity, marked by heroic-tragic ideological battles between opposing visions of what constitutes a "good" society has supposedly come to an end. Liberal political democracy is now firmly and consensually established as the uncontested and rarely examined ideal of institutionalized political life.

There are of course still ongoing rear-guard ideological battles on the geographical and political margins of the "civilized" world, waged by those who have not yet understood the lie of the land and the final horizon of history. When the need arises, they are corralled by any means necessary into consensual participation in the new global democratic order (although not always effectively, as the Afghanistan, Syria, and Iraq disasters testify). In contrast, we—the West and its allies—will now forever live happily in the complacent knowledge that democracy has been fine-tuned to assure the efficient management of a liberal and pluralist society under the uncontested aegis of a naturalized market-based configuration for the production and distribution of a cornucopia of goods and services. Any remaining problems and issues will be dealt with in the appropriate manner, through consensual forms of techno-managerial negotiation. This is supposed to be the final realization of the liberal Platonic idyll: An untroubled, undivided, cohesive, and common-sense society in which everyone knows one's place and performs one's duties in one's own (and hence in everyone's) interests, through a diversity of institutionalized forms of representative government, nurtured and supported by participatory governance arrangements for all sorts of recognized problems, issues, and matters of public concern.

Nonetheless, there is an uncanny feeling among many that all is not as it should be. Climate change keeps galloping forward despite successive impotent attempts to stem the pouring out of greenhouse gases, proliferating terrorist attacks of a variety of kind nurture a sense of perpetual insecurity, precarious jobs and highly polarized socioeconomic structures rupture the fabric of urban life, xenophobic and other ultra-identitarian movements populate the political landscape with increasing legitimacy and self-confidence, nationalist and populist movements march triumphantly in many countries. Many people feel trapped in the treadmill of pursuing their happiness through self-realization while existential angst and an uncanny sense of disempowerment haunt their nights. Some call on the powers that be to provide an immunological prophylactic that will shield them from external threats while fully recognizing that absolute immunization will only intensify the schism between the immunologized bodies of the included and the growing army of nonincluded that are increasingly descending into a form of life that Giorgio Agamben would define as "bare life" (Agamben 1998). In the meantime, the elites relentlessly pursue their self-interest and the perpetuation of the same despite increasing warnings that things are spiraling out of control. The political configuration, sclerotized as it may be, nonetheless reproduces itself as if nothing can shake the self-confidence in its own superiority. Yet, political apathy for mainstream parties and politics, and for the ritualized choreographies of representative electoral procedures, is at an all-time high. Many people have surrendered to the cynical position that nothing can and will change irrespective of one's participation in the political process. Others insist on voicing their discontent with the state of the situation by supporting extreme identitarian movements, usually of a xenophobic kind.

Nonetheless, as soon as the practices of governance were reduced to the bio-political management of the "happiness" of the population and the neoliberal organization of the transformation of nature and the appropriation and distribution of its associated wealth, new specters of the political appeared on the horizon. Insurrectional and incipient democratizing movements and mobilizations exploded in 2011, and continue to smolder and flare: Syntagma Square, Puerta del Sol, Zuccotti Park, Paternoster Square, Taksim Square, Tahrir Square, Sao Paulo, Oakland, Montreal, Hong Kong's umbrella movement, and so on. These are just a sampling of the more evocative names that have become associated with emergent new forms of politicization. Assembled under the generic banner "Real Democracy Now!," the gathered insurgents expressed an extraordinary antagonism to the instituted—and often formally democratic—forms of governing, and they staged, performed, and choreographed new configurations of the democratic. While generally articulated around an emblematic quilting point (a threatened park, devastating austerity measures, the public bailout of irresponsible financial institutions, rising tuition fees, a price hike in public transport, corrupt politicians, and the like), these movements quickly universalized their claims to embrace a desire for a transformation of the political structuring of life, against the exclusive, oligarchic, and consensual governance of an alliance of professional economic, political, and technocratic elites determined to defend the neoliberal order by any means necessary. In some case, like *Syriza* in Greece or *Podemos* in Spain, these insurgencies universalized into performative national progressive political movements intent on transforming the very parameters of political life.

In both the United States and the UK, the surprising success of self-styled socialists like Bernie Sanders's remarkable success in the Democratic Party primaries of 2016 and Jeremy Corbyn's unrelenting support from an enthusiastic popular base, all this much to the horror and disbelief of the party's establishment figures, are unmistakable signs that something is stirring politically in ways we have not seen for a long time.

It is precisely this parallax gap that sets the contours and contents of this book. From one vantage point—usually nurtured by those who seek to maintain things as they are—democracy is alive and kicking. From the other perspective, the democratic functioning of the political terrain has been eroded to such an extent that a radical reordering and reconfiguration of a "government of the people, by the people, for the people" is urgently required. A new constituent politics is clearly on the books. The latter position demands a dramatic transformation of the depoliticizing practices that have marked the past few decades. Yet these practices have only intensified in the context of a global economic crisis marked by the relentless persistence of neoliberalizing dogmas, and which staggers blindly forward in the absence or tumbling influence of its once-inspiring master discourses. Its continuity is ensured by a range of political elites from both right and left, and is legitimized by their continuous election to power—a power that has become more and more feeble as they delegate social and political choices to those demanded, staged, configured, and "imposed" by the socially disembodied "hidden hand of the market." The class divides of yesteryear seem to be replaced by the battle between nationalist and identitarian desires on the one hand and a cosmopolitan neoliberal internationalism on the other. Piercing through this deadlock between the deepening inequalities and disempowerment for some produced by the latter and the xenophobic and localist dynamics of the former is a considerable challenge for progressive and emancipatory politics, one that demands urgent attention.

The book proposes an intervention within the diverse literatures that have recently emerged around contemporary processes of post-politicization and the consolidation of post-democratic modes of governance. Already in the early 1990s, Philippe Lacoue-Labarthe and Jean-Luc Nancy provided an exquisitely dialectical exploration of what they defined as the "retreat of the political," understood as both the disappearance and the retreating of the political in both theoretical musings and modes of appearance

(Lacoue-Labarthe and Nancy 1997). A proliferating body of thought has since begun to decipher, both theoretically and empirically, the dynamics of depoliticization, and the contours and characteristics of the alleged "disappearance of the political," while considering the modalities and possibilities for a nascent "return of the political." According to this literature, contemporary forms of depoliticization are characterized by the erosion of democracy and the weakening of the public sphere as a consensual mode of governance has colonized, if not sutured, political space. In the process, agonistic political disagreement has been replaced by an ultra-politics of ethnicized and violent disavowal on the one hand, and the exclusion and containment of those who pursue a different political-economic model on the other. These extremes are placed outside the post-democratic inclusion of different opinions on anything imaginable in stakeholder arrangements of impotent participation and "good" governance. Such exiling ensures that the framework of debate and decision making does not question or disrupt the existing state of the neoliberal political-economic configuration. This process is generally referred to as one of post-politicization, institutionally configured through modes of post-democratic governance.

Articulated around the work of, among others, Chantal Mouffe, Jacques Rancière, Slavoj Žižek, Alain Badiou, Colin Crouch, and others, the book seeks to interrogate critically the forces and mechanisms through which the political—understood as a space of contestation and agonistic engagement—is being progressively colonized by politics—understood as technocratic procedures of consensual immuno-biopolitical governance that operates within a relatively unquestioned framework of representative democracy, free market economics, and cosmopolitan liberalism.

While theoretically framed within post-foundational thought, the main body of the book is grounded in urban and ecological processes, struggles, and conflicts through which post-politicizing modes of being-in-common became etched into the formal institutional and informal common-sense practices of life. Two central themes and theaters, "the city" and "nature," will be mobilized as emblematic illustrations of the construction of post-democratic modes of governance that foreclose "the political." First, the politics of neoliberal planetary urbanization and the disappearance of the urban qua "polis" will be a central theme. The second entrée into the archeology of post-politicization will focus on the socio-ecological predicament the world is in. We shall argue that "ecology functions today as the new opium of the masses" (see Badiou 2008b) and signal how the particular discursive, managerial, and political framings of Nature and the environmental condition we are in contributes to the formation of depoliticizing modes of governance. Particular attention will be paid to the impotent politics of climate change and its discontents. In the final part of the book, the spectral return of the political in the forms of insurgent practices that mark what Alain Badiou has recently called "the age of rebellion" will consider actually practiced modes of "repoliticization" visible in the interstices of urban life today (Badiou 2012). The book explores the possibilities for a reassertion of the political, and the strategies through which such possibilities are assimilated and foreclosed. Part III discusses the modalities of a progressive and emancipatory political sequence adequate to the situation we are in.

I Post-Democracy: Thinking (Post-)Politicization

The opening part of this book explores the contours of the process of post-politicization and suggests a range of theoretical avenues that attempt to make sense of the paradoxical situation the world is in. In chapter 1, I seek to explore the new forms of governance-beyond-the-state that increasingly mark public policy-making arrangements. Such arrangements are usually referred to as networked and participatory governance in which both state and private actors take part, and are generally praised as enhancing the democratic character of governance; however, the chapter interrogates their questionable democratic credentials. I conclude that the prevalence of such institutional configuration radically overhauls the state-civil society articulation, and inaugurates new forms of post-democratic governance.

This analysis sets the scene, in chapter 2, for an engagement with post-foundational political theory that may help to decipher our political predicament as a particular form of depoliticization. Marked by the dominance of techno-managerial forms of intervention that foreclose, disavow, or repress the antagonisms that cut through the social order, this form of depoliticization is defined as "post-politicization." Chapter 2 indexes the key contours of post-politicization and its manifold, often contradictory, manifestations.

Of course, all this begs the questions as to what repoliticization might entail in such context. Or in the other words, how can we think the political again in times of rapid post-politicization. This is the challenge I embark on in chapter 3. Mobilizing the conceptual difference between "politics" and the "the political," I suggest a theoretical perspective that permits thinking of depoliticization and repoliticization together as two interlinked yet profoundly different processes. The political is introduced as the immanent terrain for the expression of social antagonism and for the egalitarian staging of disagreement. The opening and nurturing of spaces for the

expression of disagreement and for experimenting with new forms of egalitarian being-in-common, I contend, is vital to political processes aimed at
inaugurating more democratic, egalitarian, and inclusive forms of doing
politics, and for the institutionalization of more egalibertarian governance
arrangements. These chapters, taken together, frame our foray into the
excavation of the environmental and urban condition that choreographs
much of the present configuration. The modalities of how post-politicizing
practices become etched in environmental and urban thought and practice
will take center stage in part II.

1 The Janus Face of Governance-beyond-the-State

Over the past few decades, a proliferating body of scholarship has attempted to theorize and substantiate empirically the emergence of new formal or informal post-democratic institutional arrangements that engage in the act of governing outside- and beyond-the-state (Rose and Miller 1992; Jessop 1998; Pagden 1998; Mitchell 2002; Hajer 2003b; Whitehead 2003; González and Healey 2005; Moulaert, Martinelli, and Swyngedouw 2006). Governance-beyond-the-state refers in this context to the emergence, embedding, and active nurturing (by the state and international bodies like the European Union or the World Bank) of institutional arrangements of "governing" that assigns a much greater role in policy making, administration, and implementation to the involvement of private economic actors on the one hand and to parts of civil society on the other in self-managing and governing what until recently was provided or organized by the national or local state. In a context of perceived or real "state-failure" on the one hand and attempts to produce systems of "good" governance on the other, institutional ensembles of governance based on such horizontally networked tripartite composition are viewed as empowering, participatory, inclusive, democracy enhancing, and more effective forms of governing compared with the sclerotic, hierarchical, and bureaucratic state forms that conducted the art of governing during much of the twentieth century. Nonetheless, while these figures of governance often offer the promise of greater participation in policy deliberation and grassroots empowerment, they also show a series of contradictory tendencies. It is precisely these tensions and contradictions that this chapter will focus on in an attempt to set the stage for considering the question of the political and the democratic in greater detail in the subsequent chapters. These historical transformations of the regimes of governance over the past few decades provide the base

characteristics that I shall define later as the process of post-democratization and post-politicization.

While much of the analysis of a changing, if not new, governmentality
(or governmental rationality (Gordon 1991) starts from the vantage point
of how the state is reorganized and mobilizes a new set of "technologies of
governing" to respond to changing socioeconomic and cultural conditions,
this chapter seeks to assess the consolidation of new forms of governance
capacity and the associated changes in governmentality (Foucault 1979) in
the context of the rekindling of the governance-civil society articulation
that is associated with the rise of a neoliberal governmental rationality and
the transformation of the technologies of neoliberal governance.

Governance as an arrangement of governing-beyond-the-state (but often
with the explicit inclusion of parts of the local, regional, or national state
apparatus) is defined as the institutional or quasi-institutional arrangements of governance that are organized as horizontal associational networks of private (market), civil society (usually NGO), and state actors
(Dingwerth 2004). These forms of apparently horizontally organized and
polycentric *ensembles* in which power is dispersed are increasingly prevalent in rule making, rule setting, and rule implementation at a variety of
geographical scales (Hajer 2003b, 175). They can be found from the local/
urban level (such as development corporations, ad hoc committees, stakeholder-based formal or informal associations dealing with social, economic,
infrastructural, environmental, or other matters) to the transnational scale
(such as the European Union, the WTO, the IMF, the free trade negotiations
[CETA, TTIP] or the successive COP Climate Conventions) (Swyngedouw
1997). They exhibit an institutional configuration based on the inclusion
of private market actors, civil society groups, and parts of the "traditional"
state apparatus. These modes of governance have been depicted as a new
form of governmentality, that is "the conduct of conduct" (Foucault 1982;
Lemke 2002), in which a particular rationality of governing is combined
with new technologies, instruments, and tactics of conducting the process
of collective rule setting, implementation, and, often, policing. However,
as Maarten Hajer argues, these arrangements take place within an "institutional void": "*There are no clear rules and norms according to which politics is
to be conducted and policy measures are to be agreed upon.* To be precise, there
are *no generally accepted* rules and norms according to which policy making
and politics is to be conducted" (Hajer 2003b, 175; emphasis in original).

The urban scale has been a pivotal terrain where these new arrangements of governance have materialized in the context of the emergence of new social movements on the one hand and transformations in the arrangements of conducting governance on the other (Le Galès 1995; Brenner and Theodore 2002; Brenner 2004; Jessop 2002c). The main objective of this chapter is to address and problematize these emerging new regimes of (urban) governance with a particular emphasis on changing political citizenship rights and entitlements on the one hand, and their democratic credentials on the other. Our focus will be on the contradictory nature of governance-beyond-the-state and, in particular, on the tension between the stated objective of deepening democracy and citizen empowerment through participation on the one hand and the often undemocratic and authoritarian character of the associated new institutional configurations on the other. This analysis is particularly pertinent as the inclusion of civil society organizations (like NGOs) in systems of (urban) governance, combined with a greater political and economic role of "local" political elites and economic actors, is customarily seen as potentially empowering and democratizing (Le Galès 2002; Hajer 2003a; Novy and Leubolt 2005). These forms of governance are innovative and often promising in terms of delivering improved collective services and they may indeed embryonically contain germs of ideas that may permit greater openness, inclusion, and empowerment of hitherto excluded or marginalized social groups. However, there are equally strong processes at work pointing in the direction of a greater autocratic governmentality (Swyngedouw 1996b, 2000; Harvey 2005) and an impoverished practice of political citizenship. These forms of governance are both actively encouraged and supported by agencies pursuing a neoliberal agenda (like the EU, the IMF, or the World Bank) and "designate the chosen terrain of operations for NGOs, social movements, and 'insurgent' planners" (see Sandercock 1998; Goonewardena and Rankin 2004). It is exactly this interplay between the empowering *gestalt* of such new governance arrangements on the one hand and their position within a broadly neoliberal political-economic order on the other that this chapter seeks to tease out.

In the first section of the chapter, we outline the contours of governance-beyond-the state. We then address the thorny issues of the state/civil society relationship in the context of the emergence of the new governmentality associated with governance-beyond-the-state. In a third section, we tease out the contradictory way in which new arrangements of governance have created new institutions and empowered new actors, while disempowering others. We argue that this shift from "government" to "governance" is associated with the consolidation of new technologies of government (Dean 1999) on the one hand, and with profound restructuring of the parameters of political democracy on the other, leading to a substantial democratic deficit and to a foreclosure of democratic spaces of agonistic encounter (Mouffe 2005). As such, this mode of governance entails a transformation both of the institutions and of the mechanisms of participation, negotiation, and conflict intermediation (Coaffee and Healey 2003). Participation, then, is one of the key terrains on which battles over the form of governance and the character of regulation are currently being fought (Raco 2000; Docherty, Goodlad, and Paddison 2001). We shall conclude by suggesting that arrangements of governance-beyond-the-state are fundamentally Janus-faced, particularly under conditions in which the democratic character of the political sphere is increasingly eroded by the encroaching imposition of market forces that set the "rules of the game" (Marquand 2004). While promising enhanced participatory embedding, governance-beyond-the-state sets the institutional configurations that we will associate in chapter 2 with more pervasive processes of post-democratization.

Governance-beyond-the-State: Networked Associations

It is now widely accepted that the system of governing in the Global North and elsewhere is undergoing rapid change (European Commission 2001; Le Galès 2002). Although the degree of change and the depth of its impact are still contested, it is beyond doubt that the nineteenth- and twentieth-century political ideal and sometimes practice of articulating the state/civil society relationship through different forms of representative democracy, which vested power in hierarchically structured transcendental state forms, is complemented by a proliferating number of new institutional forms of governing that exhibit rather different characteristics (Jessop 1995; Kooiman 1995, 2003; Grote and Gbikpi 2002; Brenner 2004). In other words, the Westphalian state-order that matured in the twentieth century in the form of the liberal-democratic state, organized at local, often also at regional, and national scales, has begun to change in important ways, resulting in new forms of governmentality, characterized by a new articulation between state-like forms (such as the European Union, urban development corporations, and the like), civil society organizations, and private market actors

(Brenner et al. 2003). While the traditional state form in liberal democracies is theoretically and practically articulated through forms of political citizenship that legitimizes state power by means of it being vested within the political voice of the citizenry, the new forms of governance exhibit a rather fundamentally different articulation between power and citizenship and constitute, according to Thomas Lemke, a new form of governmentality (Lemke 2001, 2002). As Schmitter defines it: "Governance is a method/mechanism for dealing with a broad range of problems/conflicts in which actors regularly arrive at mutually satisfactory and binding decisions by negotiating with each other and cooperating in the implementation of these decisions" (Schmitter 2002, 52).

Paquet defines governance as "the newly emerging models of action [that] result from the concerted combination of social actors coming from diverse milieus (private, public, civic) with the objective to influence systems of action in the direction of their interests" (Paquet 2001, quoted in Hamel 2002, 52; my translation).

From this perspective, it is not surprising to find that such modes of governance-beyond-the-state are resolutely put forward as presenting an idealized normative model (see Le Galès 1995; Schmitter 2000, 2002) that promises to fulfill the conditions of good government "in which the boundary between organizations and public and private sectors has become permeable" (Stoker 1998, 38). It implies a common purpose, joint action, a framework of shared values, continuous interaction and the wish to achieve collective benefits that cannot be gained by acting independently (Rakodi 2003). This model is related to a view of "governmentality" that considers the mobilization of resources (ideological, economic, cultural) from actors operating outside the state system as a vital part of democratic, efficient, and effective government (Pierre 2000a, 2000b). Schmitter continues to argue that, in a normative-idealized manner,

Governance arrangements are based on a common and distinctive set of features:

• Horizontal interaction among presumptive equal participants without distinction between their public or private status.
• Regular, iterative exchanges among a fixed set of independent but interdependent actors.
• Guaranteed (but possibly selective) access, preferably as early as possible in the decision-making cycle.
• Organized participants that represent categories of actors, not individuals. (Schmitter 2000, 4)

State-based arrangements are hierarchical and top-down (command-and-control) forms of setting rules and exercising power (but recognized as legitimate via socially agreed conventions of representation, delegation, accountability, and control), and mobilizing technologies of government involving policing, biopolitical knowledge, and bureaucratic rule. Governance-beyond-the-state systems, in contrast, are presumably horizontal, networked, and based on interactive relations between independent but interdependent actors who share a high degree of trust, despite internal conflict and oppositional agendas, within inclusive participatory institutional or organizational associations. The mobilized technologies of governance revolve around reflexive risk calculation (self-assessment), accountancy rules, and accountancy-based disciplining, quantification, and benchmarking of performance (Dean 1999).

The participants in such forms of governance partake (or are allowed to partake) in these relational networked forms of decision-making on the basis of the "stakes" they hold with respect to the issues these forms of governance attempt to address. The relevant term "stakeholder" has gained currency in recent years and propelled its associated politics of stakeholder governance to the forefront of the political platform (Newman 2001). According to Philippe Schmitter, the shift from "political citizenship" articulated through statist forms of governing to a stakeholder-based polity does not go far enough. He proposes, therefore, an enlarged approach by introducing the notion of "holder," which should constitute the foundation for establishing rights or entitlements to participate (Schmitter 2002). Table 1.1 summarizes Schmitter's extended formulation.

Of course, such an idealized-normative model of horizontal, nonexclusive, and participatory (stake)holder-based governance is symptomatically oblivious of the contradictory tensions in which these forms of governance are embedded. These new practices are indeed riddled with all manner of problems with respect to their democratic content. Arrangements are often imposed from above; there is widespread distrust, particularly as rules and norms are not agreed but decided under noncodified and often informal ad hoc principles (Hirst 1995; Dryzek 2000; Akkerman, Hajer, and Grin 2004). Before we embark on considering the democratic credentials of such post-democratic institutional ensembles, we need to turn our attention, first, to how these transformations of the conduct of conduct articulate with changing choreographies of civil society/state interaction, and, second, to

Table 1.1
Schmitter's matrix of definitions of "holders"

Right holders participate because they are members of a national political community

Space holders participate because they live somewhere affected by the policy

Knowledge holders participate because they have particular knowledge about the matter concerned

Shareholders participate because they own part of the assets that are going to be affected

Stakeholders participate because, regardless of their location or nationality, they might be affected by change

Interest holders participate on behalf of other people because they understand the issues

Status holders participate on behalf of other people because they are given a specific representative role by the authorities

---

Source: Schmitter 2000.

the emergence of these new forms of governance in the context of broader processes of political-economic regime changes.

Articulating State, Market, and Civil Society

Bob Jessop argues that the state is capitalism's necessary "other" (Jessop 2002a). For him, the social relations that produce and sustain capitalist economic forms require extra-economic rules and institutions to function. These institutions can take a variety of forms, the national liberal democratic state form that dominated the West from the late nineteenth century onward being just one of them. So, while state and market can be separated conceptually, they are functionally and strategically intimately interconnected. In addition to state and market, there is also the sum total of social forms and relations that are neither state nor market. These are usually captured under the notion of "civil society." There is considerable confusion about the status, content, and position of civil society, both analytically and empirically. This confusion arises partly from the meandering history of the concept, partly from the changing position of civil society in relation to political society (see Novy and Leubolt 2005). While the early Enlightenment view of civil society posited "civil" society versus "natural" society, Hegel and Marx considered civil society as a set of economic/material relations that operated at a distance and were different from the state and state acting. Of course, this change in perspective was in itself related to the changing nature of the state (from a "sovereign" to a biopolitical state, i.e., from a state focused on the integrity of its territorial control to one operating allegedly in the "interest of all for the benefit of all"). Liberal thinkers, like Alexis de Tocqueville, in turn, associated civil society with voluntary organizations and associations. In contrast, with Antonio Gramsci, writing at the time of the embryonic formation of the liberal-democratic Keynesian welfare state, civil society became viewed as one of the three components (the others being the state and the market) that provide content to and structures society (Gramsci 1971). For him, civil society is the collection of private actors (outside state and market) and constitutes the terrain of social struggle for hegemony (Showstack Sassoon 1987; Simon 1991; MacLeod 1999; Ekers et al. 2013). Moreover, both civil society and its meaning are also closely related to the Foucaultian notion of "governmentality."

Indeed, with the rise of the liberal state in the eighteenth century, civil society became increasingly associated with the object of state governing as well as perceived as the foundation from which the state's legitimacy was claimed. In addition, as the state turned increasingly into a biopolitical democratic state, concerned with and intervening in the "life qualities" of its citizens (health, education, crime prevention, socioeconomic well-being, among others) from whom the state draws its legitimacy through a system of pluralist democratic controls, civil society emerged as both an arena for state intervention and a collection of actors engaging with, occasionally opposing, but invariably relating to the state (Lemke 2001). At the same time, the liberal state maintained the "economic" sphere as a fundamentally "private" one, operating outside the collective sphere of the state but shaping the material conditions of civil life in a decisive manner. The social order, consequently, became increasingly understood and constructed as the articulation between state, civil society, and market. While for Hegel and Marx, albeit in very different ways, the ideal of society resided in transcending the separation between the political state and civil society, the operation of the economy, under the hidden hand of the market in liberal-capitalist societies, rendered this desired unity of state and civil society impossible. In fact, a fuzzy terrain was produced, somewhere in between yet articulating with state and market, but irreducible to either; a terrain that was neither state nor public, yet expressing a diverse set of social activities and infused with all manner of social power relations, tensions, conflicts, and social struggles. Civil society is, in other words, the pivotal terrain from which social transformative and occasionally politically innovative action emerges and where social power relations are contested and struggled over. The relative boundaries between these three instances (i.e., state, civil society, and market) vary significantly from time to time and from place to place. The notion of civil society, therefore, also cannot be understood independently of the relations between political and economic power, the first articulated in terms of access to or control over the state apparatus, the latter in terms of access to or control over resources for accumulation (in the form of monetary, material, cultural, or social capital).

In sum, the position and role of civil society is closely related to the dynamics of other "moments" of society, meaning, state and economy. At moments of increasing socioeconomic tension and restructuring (such as during the 1920s/1930s,1980s/1990s, or again after 2008), the conduct of conduct changes in such a way that continued sustained accumulation can be maintained, restored, or improved, but without undermining the relative coherence or stability of the social order. This unfolds exactly at moments when this relative coherence of society is under serious stress, showing widening ruptures and tensions, occasionally pointing at a potential disintegration of the social order itself. Successful restructuring of capitalism demands, therefore, strong governance in order to produce stronger economic dynamics (understood in market economy terms) while maintaining cohesion in civil society during periods of significant change. Such restructuring of governance indeed often takes place at exactly the time when civil society goes through painful shocks associated with that restructuring—shocks that further undermine the legitimacy of the state and reinforce calls for alternative models of governance. In other words, governing becomes more problematic and the terrains of governance begin to shift (see Poulantzas 1980a, 1980b). The state can become more authoritarian (as happens with fascism) or more autocratic, while delegating power and including new strata of civil society in the forms of governance (as is happening now) (Harvey 2005).

Foucault's notion of governmentality may help to chart recent changes in the state/civil society relationship and the emergence of arrangements of governance-beyond-the-state (Donzelot 1991; Pagden 1998). For Foucault, governmentality refers to the rationalities and tactics of governing and how they become expressed in particular technologies of governing, such as, for example, the state (Foucault 1984). The state, therefore, appears in Foucault's analysis as a "tactics of government, as a dynamic form and historic stabilization of societal power relations" (Lemke 2002, 60). Governmentality, therefore, is "at once internal and external to the state, since it is the tactics of government which make possible the continual definition and redefinition of what is within the competence of the state and what is not, the public versus the private, and so on; thus the state can only be understood in its survival and its limits on the basis of the general tactics of governmentality" (Foucault 1991, 103).

Foucault's analysis of neoliberal reason and neoliberal governmentality exactly excavates the changing role of the state in, and the reshaping of governing under, neoliberalism. Thomas Lemke summarizes the emerging new articulation between state and civil society under a neoliberal governmentality as follows:

By means of the notion of governmentality, the neo-liberal agenda for the "withdrawal of the state" can be deciphered as a technique for government. The crisis of Keynesianism and the reduction in forms of welfare-state intervention therefore lead less to the state losing powers of regulation and control (in the sense of a zero-sum game) and can instead be construed as a re-organization or re-structuring of government techniques, shifting the regulatory competence of the state onto "responsible" and "rational" individuals. Neoliberalism encourages individuals to give their lives a specific entrepreneurial form. It responds to stronger "demand" for individual scope for determination and desired autonomy by "supplying" individuals and collectives with the possibility of actively participating in the solution of specific matters and problems which had hitherto been the domain of specialized state agencies specifically empowered to undertake such tasks. This participation has a "pricetag": the individuals themselves have to assume responsibility for these activities and the possible failure thereof. (Lemke 2001, 202; see also Donzelot 1984, 157–177, Burchell 1993, 275–276)

Elsewhere, Lemke argues how a Foucaultian perspective permits a view of neoliberalism not as "the end but [as] a transformation of politics that restructures the power relations in society. What we observe today is not a diminishment or reduction of state sovereignty and planning capacities, but a displacement from formal to informal techniques of government and the appearance of new actors on the scene of government (e.g. NGOs, private organizations and economic actors, among others), that indicate fundamental transformations in statehood and a renewed relation between state and civil society actors" (Lemke 2002, 50).

This "destatization" of a series of former state domains and their transfer to civil society organizations redefines the state/civil society relationship through the formation of new forms of governance-beyond-the-state (Jessop 2002c). This involves a threefold reorganization. First is the externalization of state functions through privatization and deregulation, outsourcing, and decentralization. Both mechanisms inevitably imply that nonstate, civil society, or market-based configurations become increasingly involved in regulating, governing, organizing, and policing a series of social, economic, and cultural activities. Second is the upscaling of governance whereby the national state increasingly delegates regulatory and other tasks to other and higher scales or levels of governance (such as the EU, IMF, WTO, and the like). Third is the downscaling of governance to local practices and arrangements that create greater local differentiation combined with a desire to incorporate new social actors in the arena of governing (Swyngedouw 1997, 2004a). This includes processes of vertical decentralization toward subnational forms of governance (see Moulaert, Rodriguez, and Swyngedouw 2003). These three processes of rearrangement of the relationship between state, civil society, and market simultaneously reorganize the arrangements of governance as new institutional forms of governance-beyond-the-state are established and become part of the system of governing, of organizing the conduct of conduct. This restructuring is embedded in a consolidating neoliberalizing polity. The latter combines a desire to politically construct the market as the preferred social institution of resource mobilization and allocation, a critique of the "excess" of state associated with Keynesian welfarism, and engineering of the social in the direction of greater individualized responsibility (Peck 2010). Of course, this scalar reorganization of the state and the associated emergence of a neoliberal governance-beyond-the-state redefine in fundamental ways the state/civil society relationship. The new articulations between state, market, and civil society generate new forms of governance that combine the three "moments" of society in new and often innovative ways (Swyngedouw 2004a).

Of course, the new modalities of governance also involve the mobilization, by the state, of a new set of technologies of power, which Mitchell Dean identifies as technologies of agency and technologies of performance (Dean 1999). While the former refers to strategies of rendering the individual actor responsible for his or her own actions, the latter refers to the mobilization of benchmarking rules that are set as state-imposed parameters against

which (self-)assessment can take place and that require the conduct of a
particular set of performances that can be monitored, quantified, assessed,
and transmitted. These technologies of performance produce "calculating
individuals" within "calculable spaces" and incorporated within "calculative
regimes" (Miller 1992). Barbara Cruikshank refers in this context to the mobilization of "technologies of citizenship" (Cruikshank 1993, 1994), which are
defined as "the multiple techniques of self-esteem, of empowerment, and
of consultation and negotiation that are used in activities as diverse as community development, social and environmental impact assessment, health
promotion campaigns, teaching at all levels, policing, research, the combat-
ing of various kinds of dependency and so on" (Dean 1999, 168). Ironically,
while these technologies are often advocated and mobilized by NGOs and
other civil organizations speaking for the disempowered or socially excluded
(Carothers, Brandt, and Al-Sayyid 2000), these actors often fail to discern how
these instruments are an integral part of the consolidation of an imposed
and authoritarian neoliberalism, celebrating the virtues of self-managed
risk, prudence, critical reflexivity with respect to the benchmarks, and self-
responsibility (Castel 1991; O'Malley 1992; Burchell 1996; Dean 1995, 1999).

To the extent that *participation* is invariably mediated by *power* (whether
political, economic, gender, or cultural) among participating "holders,"
between scales of governance/government and between governing institutions, civil society, and encroaching market power, the analysis and understanding of shifting relations of power is a central concern (see Getimis
and Kafkalas 2002). Since it is impossible within the remit of this chapter
to review exhaustively the possible theorizations and perspectives on social
and political power, we focus on the principles that fundamentally shape
individuals' or social groups' position within the polity and that articulate
their respective (but interrelated) power positions vis-à-vis governing institutions on the one hand, and within civil society on the other. In particular,
in what follows, we take the theoretical and practical yardsticks of what
constitutes democratic government together with the practices associated
with arrangements of governance-beyond-the-state.

The Democratic Deficit of Governance-beyond-the-State: Toward Post-Democratization

While in pluralist instituted democracy, the political entitlement of the
citizen is articulated via the twin condition of "national" citizenship on

the one hand, and the entitlement to political participation in a variety of ways (but, primarily via a form of constitutionally or otherwise-codified representational democracy) on the other, network-based forms of governance do not (yet) have codified rules and regulations that shape or define participation and identify the exact domains or arenas of power (Hajer 2003b). As Ulrich Beck argues, these practices are full of "unauthorized actors" (Beck 1999, 41). While such absence of codification potentially permits and elicits socially innovative forms of organization and of governing, it also opens up a vast terrain of contestation and potential conflict that revolves around the exercise of (or the capacity to exercise) entitlements and institutional power. The status, inclusion or exclusion, legitimacy, system of representation, scale of operation, and internal or external accountability of such groups or individuals often take place in nontransparent, ad hoc, and context-dependent ways and differ greatly from those associated with pluralist democratic rules and codes. While the democratic lacunae of pluralist liberal democracy are well known, the procedures of democratic governing are formally codified, transparent, and usually easily legible. The modus operandi of networked associations is much less clear. Moreover, the internal power choreography of systems of governance-beyond-the-state is customarily led by coalitions of economic, socio-cultural, or political elites (Swyngedouw, Moulaert, and Rodriguez 2002). Therefore, the rescaling of policy transforms existing power geometries, resulting in a new constellation of governance articulated via a proliferating maze of opaque networks, fuzzy institutional arrangements, ill-defined responsibilities, and ambiguous political objectives and priorities. In fact, it is the state that plays a pivotal and often autocratic role in transferring competencies (and consequently for instantiating the resulting changing power geometries) and in arranging these new networked forms of governance. The democratic fallacies of the pluralist democratic state are compounded by the expansion of the realm of governing through the proliferation of such asymmetric governance-beyond-the-state arrangements. In fact, when assessing the formal requirements of pluralist democracy against the modes of arrangements of governance-beyond-the-state, the contradictory configurations of these networked associations come to the fore and show the possible perverse effects or, at least, the contradictory character of many of these shifts. We will examine six key themes associated with pluralist democratic governance and how they change in the process of post-democratization. These six markers are entitlement and status, the structure of representation, accountability, legitimacy, scales of governance, and order of governance.

Entitlement and Status

The first theme revolves around the concepts of "entitlement" and "status". While the concept of (stake-)holder is inclusive and presumably exhaustive, the actual concrete forms of governance are necessarily constrained and limited in terms of who can participate, is participating, or will be allowed to participate. Hence, status and assigning or appropriating entitlement to participate are of prime importance. In particular, assigning holder status to an individual or social group is not neutral in terms of exercising power. In most cases, entitlements are conferred upon participants by those who already hold a certain power or status. Of course, the degree to which mobilizations of this kind are successful depends, among other factors, on the degree of force or power, or both, that such groups or individuals can garner, and on the willingness of the existing participants to agree to include them. In addition, the terms of participation may vary significantly from mere consultation to the right to vote and co-decide. Needless to say, status within the participatory rituals co-determines effective power positionality. More fundamentally, while political citizenship-based entitlements are (formally) inclusive (at least at a national level) and are based on a "one person one vote" rule, holder entitlements are invariably predicated upon willingness to accept groups as participants on the one hand, but also on willingness to participate on the other. The latter of course depends crucially on the perceived or real position of power that will be accorded to incumbent participants. In a context in which, partly through the erosion of political power (compared with other forms of power) and partly through an emerging more problematic relationship between state and civil society, many individuals and social groups have fully or partially "opted-out" of political participation and chosen either other forms of more direct political action or plain rejection of participation. Deep ecologists, part of the alternative globalization and anti-capitalist movements and even segments of the "social economy" sectors, have gone in this direction (Hertz 2002).

The Structure of Representation

The second theme highlights how in addition to decisions over entitlement to participate, the structure of representation is of crucial importance.

While pluralist democratic systems exhibit clear and mutually agreed forms of representation, "holder" participation suffers from an ill-defined and diffused notion of an actual system of representation (Edwards 2002). Various groups and individuals participating in networks of governance have widely diverging mechanisms of deciding on representation and organizing feedback to their constituencies. To the extent that it is primarily civil society organizations that participate in governance, their alleged insertion into grassroots civil society power is much more tenuous than is generally assumed. In fact, it proves to be extremely difficult to disentangle the lines of representation (and mechanisms of consultation and accountability that are directly related to the form of representation) through which groups (or individuals) claim entitlement to holder status (and, hence, to participation) or are assigned holder status. This, of course, opens up a space of power for the effective participants within the organization that is not at all, or only obliquely, checked by clear lineages of representation.

Accountability

The third theme, and directly related to the structure of representation, focuses on the mechanisms and lineages of accountability that are radically redrawn in arrangements of governance-beyond-the-state (Rhodes 1999; Rakodi 2003). Again, while a democratic polity has more or less clear mechanisms for establishing accountability, holder representation fundamentally lacks explicit lines of accountability. In fact, accountability is assumed to be internalized within the participating groups through their insertion into (particular segments) of civil society (through which their holder status is defined and legitimized). However, given the diffuse and opaque systems of representation, accountability is generally very poorly developed, if at all. In other words, effective representation has to be assumed, is difficult to verify, and practically impossible to challenge. The combined outcome of this leads to often more autocratic, nontransparent systems of governance that—as institutions—wield considerable power and, thus, assign considerable, albeit internally uneven power, to those who are entitled (through a selective and often random process of invitation) to participate.

Legitimacy

This brings the argument directly to the centrality of legitimation. The mechanisms of legitimation of policies or regulatory interventions or both become very different from those of representational pluralist democracy. To the extent that legitimation does not result from the organization of entitlement, representation, and accountability, these news forms of governance face considerable internal and external problems with respect to establishing legitimacy. In fact, this has been a long-running problem for many of the new forms of governance, particularly as coercion and the legitimate use of coercive technologies and violent imposition lie largely, albeit by no means exclusively, with the state. Legitimacy depends, therefore, more crucially on the linguistic coding of the problems and strategies of action. This is particularly pertinent in a policy environment that, at the best of times, only reflects a partial representation of civil society. As Kooiman notes, governance implies "a linguistic coding of problem definitions and patterns of action" (cited in Grote and Gbikpi 2002, 13). This view parallels recent theories of political consensus formation (see chapter 2), which imply a reliance on the formation of discursive constructions (through the mobilization of discourse alliances) that produces an image, if not an ideology, a representation of a desirable good, while, at the same time, ignoring or silencing alternatives and often irrespective of the factual accuracy of the argument. Post-truth discourses are indeed a key characteristic of post-democratic forms of governance. These discursive or representational strategies have become powerful mechanisms for producing hegemony and, with it, a shaky legitimacy. The latter, of course, remains extremely fragile, as it can be continuously undermined by means of counter-hegemonic discourses and the mobilization of a deconstructionist apparatus for deciphering the coding of power that is imbedded in legitimizing discourses.

Scales of Governance

The fifth theme addresses the geographical scale at which forms of governance-beyond-the-state are constituted and their internal and external relational choreographies of participation/exclusion are clearly significant. When governance-beyond-the-state involves processes of "jumping scales" (Smith 1984)—that means the transfer of policy domains to subnational or transnational forms of governance—the choreography of actors changes as well. As Maarten Hajer contends, scale jumping is a vital strategy to gain power or influence in a multi-scalar relational organization of networks of governance (Hajer 2003b, 179). For example, where national urban policy increasingly replaced local public-private partnerships, the types of social actors and their positions within the geometries of power changed as well. In other words, upscaling or downscaling is not socially neutral as new actors emerge and consolidate their position in the process, while others are excluded or become more marginal. In sum, with changing scalar configurations, new groups of participants enter the frame of governance or reinforce their power position, while others become or remain excluded.

Orders of Governance

Finally, as both Kooiman and Jessop attest, distinctions, at least theoretically, have to be made between meta-governance and first-order and second-order governance (Kooiman 2000; Jessop 2002b). Meta-governance refers to the institutions or arrangements of governance where the grand principles of governmentality are defined (Whitehead 2003). For example, the European Union, the World Trade Organization, or the G-8 meetings are textbook examples of vehicles of meta-governance. First-order governance is associated with codifying and formalizing these principles, while second-order governance refers to the sphere of actual implementation of concrete policies. In terms of political and social framing of policies, there is a clear hierarchy between these orders of governance, which can and do operate at all spatial scales. However, the choreography of participation, including entitlement, status, and accountability, varies significantly depending on the "order" of the governing network.

The Janus Face of Governance-beyond-the-State: The Contradictions of Post-Democratic Governance

Of course, the political and institutional constellation does not operate independently of the social and economic sphere. In fact, any operation of the political sphere is, de facto, a political-economic intervention as governance inevitably impinges on decisions over economic processes and modalities of use and transformation of social and environmental resources. This is particularly true in a market economy, in which key decisions over resource allocation, use, and transformation are taken by private actors who operate within the constraining or enabling regulatory framework of systems of governance. To the extent that over the past few decades there has been a tendency toward deregulation and reregulation and toward the externalization of state functions, the emergence of new forms of governance was either instrumental in shaping this transformation or these forms became established as the regulatory framework for managing a beyond-the-state polity. In this sense, the power geometries within and between networks of governance as well as, most importantly, the theater for their operation and focus of their intervention, are shaped by these wider political-economic transformations.

It would of course be premature to announce the death of the state in the wake of the emergence of these new forms of governance. In fact, many of these networked organizations are both set up, and directly or indirectly controlled, by the state, and, regardless of their origins, necessarily articulate with the state. Hence, the political power choreography in this hybrid government/governance configuration is multi-layered, diffuse, decentered, and ultimately not very transparent. Yet, whether we are considering EU levels of governance, or the emergence of subnational scales of governance (social economy initiatives, development corporations, local social movements), these cannot operate outside or independent of the state. However, their institutional operation beyond-the-state permits, in fact, a form of governmentality that is only apparently outside the state, and to which the state must necessarily respond. This ambiguity becomes one of the means the state mobilizes to deal with its own immanent legitimation crisis. For example, the new forms of governance (at the EU, IMF, or other scales, for example) are invoked by the state to legitimize and push through forms of intervention that might otherwise meet with considerable resistance from (significant parts of) civil society. The imposition of the 3 percent budget norm on national governments by the Maastricht treaty in the run-up to European monetary integration, for example, was a classic example of this practice, as is the imposition of strict austerity rules on crisis-stricken Greece, Spain, or Portugal after 2010. In the absence of clear channels of representation and accountability, civil society individuals and groups find it more difficult to engage in public debate and to contest or change courses of action decided beyond-the-state.

Therefore, the thesis of the transition in socioeconomic regulation from statist command and control systems to horizontal networked forms of participatory governance has to be qualified in a number of ways. First of all, the national or local state and its forms of political-institutional organization and articulation with society remain important. In fact, the state takes center stage in the formation of the new institutional and regulatory configurations associated with governance (Swyngedouw, Moulaert, and Rodriguez 2002). This configuration is directly related to the conditions and requirements of neoliberal governmentality in the context of a greater role of both private economic agents as well as more vocal civil society-based groups. The result is a complex hybrid form of government/governance (Bellamy and Warleigh 2001).

Second, the non-normative and socially innovative models of governance as nonhierarchical, networked, and (selectively) inclusive forms of governmentality cannot be sustained uncritically. While governance promises and on occasion delivers a new relationship between the act of governing and society, and thus rearticulates (Moulaert, Rodriguez, and Swyngedouw 2003) immanent forms of governing on the one hand, and the imposition of a transcendental Hobbesian leviathan on the other, there are also significant counter-tendencies. In particular, as discussed previously, tensions arise between:

(a) The possibilities and promises of enhanced democratization through participatory governance versus the actualities of nonrepresentational forms of autocratic elite technocracy.

(b) The extension of holder participation as partially realized in some new forms of governance versus the consolidation of beyond-the-state arenas of power-based interest intermediation.

(c) The improved transparency associated with horizontal networked inter-dependencies versus the gray accountability of hierarchically articulated and nonformalized and procedurally legitimized associations of governance.

These tensions are particularly prevalent and acute in the context of the processes of re-scaling of governance. The upscaling, downscaling, and externalization of functions traditionally associated with the scale of the national state have resulted in the formation of institutions and practices of governance that all express the preceding contradictions. This is clearly evident in the context of the formation (and implementation) of a wide array of new urban and local development initiatives and experiments on the one hand, and in the construction of the necessary institutional and regulatory infrastructure that accompanies such processes on the other (Moulaert, Rodriquez, and Swyngedouw 2003). Needless to say, this ambiguous shift from government to a hybrid form of government/governance, combined with the emergence of a new hierarchically nested and relationally articulated "gestalt of scale," constitutes an important and far-reaching socio-political transformation.

Third, the processes of constructing these new choreographies of governance are associated with the rise to prominence of new social actors, the consolidation of the presence of others, the exclusion or diminished power position of groups that were present in earlier forms of government, and the continuing exclusion of other social actors who have never been included. The new gestalt of scale of governance has undoubtedly given a greater voice and power to some organizations (of a particular kind, i.e., those who accept playing according to the rules set from within the leading elite networks). However, it has also consolidated and enhanced the power of groups associated with the drive toward marketization, and diminished the participatory status of groups associated with social-democratic or anti-privatization strategies.

Finally, and perhaps most importantly, governance-beyond-the-state is embedded within autocratic modes of governing that mobilize technologies of performance and of agency as a means of disciplining forms of operation within an overall program of responsibilization, individuation, calculation, and pluralist fragmentation. The figures of horizontally organized stakeholder arrangements of governance that appear to empower civil society in the face of an apparently overcrowded and excessive state are, in the end, not only a Trojan Horse that diffuses and consolidates the market as the principle institutional form, but even more importantly, signals a profound and disturbing transformation of the registers of governing in the direction of post-democratic forms of running and managing contested public affairs. The contours and dynamics of this transformation will be taken up in chapter 2.

2 Interrogating Post-Democratization: Post-Politicization as Techno-Managerial Governance

In fact, historical faith has changed camps. Today's faith seems to be the prerogative of governors and their experts. ... Proclaiming themselves to be simply administrating the local consequences of global historical necessity, our governments take great care to banish the democratic supplement. Through the invention of supra-State institutions which are not States, which are not accountable to any people, they realize the immanent ends of their very practice: depoliticize political matters, reserve them for places that are non-places, places that do not leave any space for the democratic invention of polemic. So the State and their experts can quietly agree amongst themselves. (Rancière 2006a, 81–82)

There is a shift from the model of the *polis* founded on a center, that is, a public center or *agora*, to a new metropolitan spatialization that is certainly invested in a process of depoliticization, which results in a strange zone where it is impossible to decide what is private and what is public. (Agamben 2006)

Although democracy is today firmly and consensually established as the uncontested and rarely examined ideal form of institutionalized political life, its practices seem to be increasingly reduced to the public management of what Alain Badiou calls the "capitalo-parliamentary order" (Badiou 2008c), a pluralist political framework that revolves around securing the sustainability of a capitalist and a broadly neoliberal socioeconomic order. An emerging body of thought has indeed begun to consider the suturing of the political by a consensual mode of governance that has apparently reduced political conflict and disagreement over possible futures either to an ultra-politics of radical and violent disavowal, exclusion, and containment or to a para-political inclusion of different opinions on anything imaginable (as long as it does not question fundamentally the existing state of the neoliberal political-economic configuration) in arrangements of impotent participation and consensual "good" techno-managerial governance (see, among others, Žižek 1999; Crouch 2000; Marquand 2004; Brown 2005; Mouffe 2005; Purcell

2008; Rosanvallon 2008; Agamben et al. 2009; Dean 2009, Hermet 2009, Swyngedouw 2011). However, this consensualism in policing public affairs is paralleled by all manner of often-violent insurgent activism and proliferating manifestations of discontent, exemplified by the success of both radical progressive and xenophobic-nationalist-identitarian movements.

I argue in this and chapter 3 that considering the irreducible difference between politics/policy-making (*la politique*) and the political (*le politique*) might open up a promising avenue for interrogating "the return of the political" as manifested by these various insurgent movements. While *la politique* refers to the interplay of social, political, and other power relations in shaping everyday policies and managerial procedures within the given instituted organizational order, *le politique* is discernable in the immanence of spaces and activities that aspire to the arrangement of egalitarian public encounter of heterogeneous groups and individuals (Marchart 2007) and in the performative staging and acting of equality in the face of the inegalitarian practices embodied in the instituted democratic order.

This sense of the political requires urgent attention again as a "relatively autonomous" (to use Louis Althusser's proposition) field. For Alain Badiou, for example, socioeconomic "analysis and politics are absolutely disconnected": the former is a matter for expertise and implies hierarchy; the latter is not. An absolute separation has to be maintained, he argues, between "science and politics, analytic description and political prescription" (Badiou 1998, 2; see also Hallward 2003a). For Badiou, the political is not a reflection of and related in an immediate sense to something else, like the cultural, the social, or the economic. Instead, it is the affirmation of the capacity of each and every one to act politically within a given "state of the situation." It is a site open for occupation by those who call it into being, render it visible, and stage its occupation, irrespective of the place they occupy within the social edifice. Badiou insists on the political as an emergent property, articulated through "an event" understood as an immanent interruption in the state of the situation. A political sequence—a process of politicization—unfolds through a process of universalization of the declaration of fidelity to the egalitarian truth expressed in the inaugural event. An event can therefore only be discerned as political retroactively, when the truth of the new situation has been established. Symbolizations of political events—like the proper names "October 1917" or "The Paris Commune"—only take place ex-post; they cannot be named in advance. It is the work of the political that inaugurates and names the truth of the event and forges symbolizations, practices, and contents, and that provides substance to the political sequence. It is not philosophy or (critical) social theory that founds or inaugurates the political. On the contrary, the political emerges through the process of political subjectivation, the process of immanent appearance in public of those who disrupt the state of the situation in the name of "equality," "the people," the "common," and "the democratic."

In this chapter, I seek to interrogate the political as understood and set forth earlier. While considerable intellectual effort has gone into excavating the practices of actually instituted policies and politics (see chapter 1), very little attention is paid to what constitutes political democracy as a political configuration associated with a particular public space (Marchart 2007). The aim of this and the subsequent chapter is to recenter political thought and practice again by exploring the views of a series of political philosophers and interlocutors who share the view that the political needs urgent attention, particularly in a discursive context that is sutured by a view of the "end of politics" and intensifying practices of depoliticization. This is not a claim to a "new" theorization of what "the political" really is. Rather, I propose an intellectual foray into grappling with the emergence in public spaces of heterogeneous, but decidedly political, rebellious, and insurrectional movements whereby the process of subjectivation, the act of participating in the politicizing event, operates at a distance from both state and theory. This perspective focuses on the political as appearance, as an immanent practice (see Arendt 1973; Dikeç 2015). Of course, this perspective also does not downplay or ignore the central importance of instituted politics, in other words, the rituals and choreographies of everyday, formally codified and organized policy-making. The latter is really the domain of actually existing and observable forms of conflict intermediation and policy formulation, implementation, and policing. However, these forms of institutionalization are interdependent with the process of politicization understood as the reassertion of the political as agonistic encounter within public space. In other words, the political both affirms the impossibility of a fully sutured society and opens up possibilities and trajectories for emancipatory change within the instituted order.

I shall first consider the process of post-politicization as a concrete and specific tactic of depoliticization, understood as the reduction, accelerating rapidly over the past few decades, of the political terrain to a post-democratic arrangement of oligarchic policing. The latter refers to the domination, to the attempted suturing or colonization of social space, by an instituted police order in which expert administration, the consensual framing of agreed-upon problems as posing techno-managerial challenges, the naturalization of politics to the management of a presumably inevitable socioeconomic ordering, and the desire for "good governance" by an administrative elite in tandem with an economic oligarchy have occupied and increasingly attempt to fill out the spatiality of the political. In other words, the space of the political is increasingly colonized, and sutured, by the spaces of the police/policies. In chapter 3, I shall attempt to recenter the political by drawing on the work of a range of political theorists and philosophers who have begun to question this post-politicizing process. Despite significant differences among them, they share a series of common understandings about what constitutes the domain of the political as a mode of active intervention in "the state of the situation" through performatively staging equality.

Post-Politicization as Depoliticization

In recent years, political scientists have lamented the decline of the public sphere, the "retreat of the political" or "the colonization of the political by the social" (Lacoue-Labarthe and Nancy 1997). David Marquand, for example, argued how the public domain of citizenship has been under attack in the United Kingdom, "first from the market fundamentalists of the New Right, and then from their New Labour imitators, resulting in a hollowing out of citizenship, the marketization of the public sector; the soul-destroying targets and audits that go with it; the denigration of professionalism and the professional ethic; and the erosion of public trust" (Marquand 2004,172). While the formal envelope of democracy survives, "its substance is becoming ever more attenuated" (ibid., 4). Pierre Rosanvallon insists how politics is increasingly substituted "by widely disseminated techniques of management, leaving room for one sole actor on the scene: international society, uniting under the same banner the champions of the market and the prophets of the law" (Rosanvallon 2006, 228). The erosion of democracy has been noted in geography too, particularly in debates over the privatization of public space (Purcell 2008; Staeheli and Mitchell 2008), the transformation of the spatialities of public encounter (Barnett 2004), the heterogeneity of struggles over the private/public nexus (Low and Smith 2005), and possible strategies for recapturing space for emancipatory purposes (Featherstone 2008, Staeheli 2008, Springer 2010, Dikeç 2017)

Pierre Rosanvallon argues that the preoccupation with depoliticized and managerial "good" governance relates to those who embrace "the development of a new type of civil society that would finally substitute for the world of politics. On this front one finds the naïve representatives of NGOs—leftists who have reinvented themselves as humanitarians—and the executives of multinational corporations, all of whom commune together today in a touching defense of an international civil society. The utopias of the one, alas, are hardly different from the hypocrisies of the others" (Rosanvallon 2006, 228).

Slavoj Žižek sees such governance as an integral part of a process of post-politicization, whereby politics is increasingly replaced by hard and soft technologies of administration (policing) of environmental, social, economic, or other domains (Žižek 2002a, 303). This post-political frame reduces politics to the sphere of governing and policy-making through allegedly participatory deliberative procedures of governance-beyond-the-state, with a given distribution of places and functions, one that excludes those who are deemed "irresponsible." It is policy-making set within a given distribution of what is possible or acceptable and driven by a desire for consent within a context of recognized difference. The stakeholders (i.e., those with recognized right to speech and right to be heard) are known in advance and disruption or dissent is reduced to the choreographies of instituted and institutional modalities of governing, the technologies of expert administration and management, to the dispositives (see Agamben 2007) of good governance within spaces appropriate for their enactment, like parliaments, council chambers, community centers, public-private governance arrangements, and so on.

This process takes the scandalous proposition of Marx that the state is the executive branch of the capitalist class as literally true: identifying politics with the management of capitalism and its contradictions is no longer a hidden secret behind the appearance of formal democracy; it has become the openly declared basis for democratic legitimacy. Maximizing the enjoyment of the people can only be achieved by declaring the inability or incapacity of "the People" (as a political name) to arrange or manage themselves the conditions of this maximization. The power of post-democracy resides,

in other words, "in the declaration of the people's impotence to act politically" (Rancière 1998, 113). Moreover, any denunciation or any struggle
against this tactic of depoliticization is regarded as going against historical
necessity. As Alain Badiou argues, "since it is commonly held that Marxism consists in assigning a determining role to the economy and the social
contradictions that derive from it, who isn't Marxist today? The foremost
'Marxists' are our masters, who tremble and gather by night as soon as the
stock market wobbles or the growth rate dips" (Badiou 2012, 7–8). Indeed,
the recent imposition of extraordinary austerity regimes, for example, by
the allied national, European, and global elites on the weakest segments of
the Spanish, Greek, Irish, or Portuguese populations precisely signals the
overwhelming, naturalized, and apparently inescapably performative effect
of economic forces and processes in whose name and integrity an outright
class war is fought (Varoufakis 2017). Even if the bullwhip of the IMF or the
European Central Bank is not mobilized, governments systematically pursue a singular and single-minded policy irrespective of the political composition of their majorities. Consider how, for example, the British, Dutch, or
Danish governments, and many others, doggedly pursue radical austerity
policies in the name of saving civilization as we know it, even if both evidence shows, and leading economic observers like Paul Krugman or Joseph
Stiglitz systematically abhor, the social and economic ravages their radical
class politics inflict while making things worse for most people. But this
is a class war fought by experts, consultants, "economists," and assorted
other elite bureaucrats and policy-makers, in close consultation with business elites, and presumably socially disembodied "financial markets." As
Marx long ago already asserted, class is a bourgeois practice and concept.
In contrast to this, the insurgencies against this class project are not waged
by a class, but by the masses as an assemblage of heterogeneous political
subjects. It is when the masses as a political category stage their presence
that the elites recoil in horror.

    The voices of those who disagree are deemed wrong, nonsensical, inar-
ticulate, and incomprehensible. They are registered as "noise" devoid of a
proper political "voice." They are not part of the distribution of the sensible. This split between noise and voice is not just a question of mutual
incomprehension that can be managed through a Habermasian open communicative act, but rather an expression of a *mésentente* (the original title
of Rancière's book on politics, problematically translated into English as

"disagreement"), a misunderstanding that is constitutive; it is a process of fundamentally rendering some voices incomprehensible and nonsensical, reducing those who disagree to the political margins and rendering them politically mute and inexistent (see also Badiou 2012).

Post-political managerialism elevates the "scandal of democracy" to new heights. This scandal refers to the democratic promise of the identity of the state with the people, a promise that must, of necessity, annul the constitutive antagonisms that cut through "the people." Those who occupy (temporarily) the place of power in democracy must suture the social order, contain the inherent antagonisms of the social order, reduce other claims to speaking as "the people" to unarticulated and nonsensical noise, by claiming they speak in the name of the People as undivided One. It is precisely within this aporetic space that Claude Lefort (but see also Hannah Arendt [1973] from a slightly different perspective) locates the totalitarian kernel of democratic forms (Lefort 1986). Instituted democracy's dark underbelly resides precisely in how its identification with the population can drive toward a position whereby the occupation of the place of power identifies with the whole of the people, disavows the constitutive conflicts within the social order and the gap between the place of politics and the social ordering of the people. Post-politicization is caught in this tension: the disappearance of the political as the space for the enunciation of dissensus (discussion follows) and the suturing of social space by a consensualizing post-political order harbors authoritarian gestures, precisely by foreclosing, disavowing, or repressing the political. Consensus politics, in a very precise sense, is for Rancière the key condition of the contemporary configuration of public administration, governance, and the arrangement of social life (Rancière 2003a, §4–6).

Post-Democracy

Colin Crouch, Jacques Rancière, and Chantal Mouffe associate the political form of this consensus politics with the emergence and deepening of post-democratic institutional configurations (Rancière 1996, 1998; Crouch 2000, 2004; Mouffe 2005):

Postdemocracy is the government practice and conceptual legitimation of a democracy *after the demos*, a democracy that has eliminated the appearance, miscount, and dispute of the people and is thereby reducible to the sole interplay of state mechanisms and combinations of social energies and interests. ... Consensus democracy is a reasonable agreement between individuals and social groups who have understood that knowing what is possible and negotiating between partners are a way for each party to obtain the optimal share that the objective givens of the situation allow them to hope for and which is preferable to conflict. But for parties to opt for discussion rather than a fight, they must exist as parties. ... What consensus thus presupposes is the disappearance of any gap between a party to a dispute and a part of society. ... It is, in a word, the disappearance of politics. (Rancière 1998, 102; see also Mouffe 2005, 29)

This arrangement assumes that "all parties are known in a world in which everything is on show, in which parties are counted with none left over and in which everything can be solved by objectifying problems" (Rancière 1998, 102). There is no excess left over and above that what is instituted. The irreducible gap between politics as the instituted order of and for the people on the one hand and, on the other, the political as the terrain of dissensual dispute articulated by those who do not count, who are surplus to the count of the situation, becomes foreclosed, repressed, or disavowed through this procedure (see Wilson and Swyngedouw 2015).

For Colin Crouch, post-democratic institutions signal a significant decline of government by the people and for the people. Although the formal architecture of democracy is still intact, there is a proliferating arsenal of new processes that bypass, evacuate, or articulate with these formal institutions (Crouch 2000, 2004). In a landmark publication, Colin Crouch defined this emerging new regime as "post-democracy," a condition he describes as follows:

While elections certainly exist and can change governments, public electoral debate is a tightly controlled spectacle, managed by rival teams of professional experts in the techniques of persuasion, and considering a small range of issues selected by those teams. The mass of citizens plays a passive, quiescent, even apathetic part, responding only to the signals given them. Behind the spectacle of the electoral game, politics is really shaped in private by interaction between elected governments and elites that overwhelmingly represent business interests. ... Under the conditions of a post-democracy that increasingly cedes power to business lobbies, there is little hope for an agenda of strong egalitarian policies for the redistribution of power and wealth, or for the restraint of powerful interests." (Crouch 2004, 4)

Richard Rorty associated post-democracy with the rapid erosion of democratic rights and values, and offers an even more chilling vision: "At the end of this process of erosion, democracy would have been replaced by something quite different. This would probably be neither military dictatorship nor Orwellian totalitarianism, but rather a relatively benevolent despotism, imposed by what would gradually become a hereditary nomenklatura" (Rorty 2004). Jacques Rancière defines post-democracy as consensus democracy, "a political idyll of achieving the common good by an enlightened government of elites buoyed by the confidence of the masses" (Rancière 1998, 93). For him, this post-democratic order revolves around a consensual arrangement in which all those who are named and counted participate within a given and generally commonsensically accepted and shared/partitioned social and spatial distribution of things and people. It is the imposition of a Platonic view whereby everyone knows their place and role within the distribution of the sensible, within the social order, and acts accordingly. While there may be recognized conflicts of interest and differences of opinion, there is widespread agreement over the conditions that exist and what needs to be done (Rancière 2003b, 2).

Geographers and scholars in cognate disciplines have begun to argue and show how post-democratization processes unfold in and through socio-spatial, environmental, and scalar transformations. Mustafa Dikeç (2007), Kevin Ward (2007), Guy Baeten (2009), Ronan Paddison (2009), Alan Cochrane (2010), Phil Allmendinger and Graham Haughton (Allmendinger and Haughton 2010), Gordon MacLeod (2011), Mike Raco (2012), Ingolfur Blühdorn (2013), and Anneleen Kenis and Erik Mathijs (Kenis and Mathijs 2014), among others, documented how recent transformations in urban or environmental governance dynamics mark the emergence of consensual modes of policy-making within new institutional configurations articulated around public-private partnerships operating in a frame of generally agreed objectives (such as sustainability, competitiveness, responsibility, resilience, inclusion, participation). I have argued how environmental debate and policing restructures spaces of governmentality in post-democratic directions (Swyngedouw 2007b, 2009a, 2015; Oosterlynck and Swyngedouw 2010; see also chapters 4 and 5), while Jim Glassman has advanced similar arguments in his work on Thailand (Glassman 2007). Some have also argued how the flip side of post-democratization forecloses politicization such that outbursts of violence remain one of the few options left to express and stage discontent and dissensus. Urban violence, in particular, has been foregrounded as a socio-spatial marker of post-politicization (Diken and Laustsen 2004; Dikeç 2007, 2017; Kaulinfreks 2008; Žižek 2008c; Swyngedouw 2017a).

The emergence of this process of post-democratization combines a series of interrelated dynamics that form a diffuse set of practices and take different forms in different places, but share a range of uncanny similarities. I will briefly enumerate its most salient characteristics.

Post-Democratization

The Economization of Politics
From the outset the political process of neo-liberalization, despite its heterogeneous, differentiated, and uneven dynamics (see Harvey 2005; Peck, Theodore, and Brenner 2009), has been marked by what Bronwen Morgan called an "economization of politics" and Wendy Brown called the "economization of everything," a process by which only those public political choices are deemed reasonable that can be incorporated within a strict market logic (Morgan 2003; Linhardt and Muniesa 2011; Brown 2015). A particular fantasy of autopoietic organization of "the economy" has indeed sutured political imaginaries of how to produce and organize the distribution of social wealth, one centered on a practice that seemingly separates economic dynamics from the political process. At the same time, much of the concern regarding governmental policy efforts is that they are geared for assuring the "proper" functioning of this fantasy in the real movement of economic life, often despite recurrent evidence of absent economic self-regulation and stabilization. The recent history of bank bailouts and variegated attempts to restore financial order in the midst of profound crisis is a case in point. This depoliticization of the economy limits or circumscribes the political choices offered to the citizen, something Henri Giroux referred to as the "terror of neoliberalism" (Giroux 2004; Purcell 2008). Moreover, the available options are often deemed too complex for ordinary citizens (or even professional politicians) to comprehend or to judge, which necessitates continuous appeal to experts to legitimize decisions (Sloterdijk 2005). The growing apathy of ordinary people with respect to the democratic political process is noted but minimized as not central to the "proper" functioning of democratic institutions (Vergopoulis 2001).

This process is a profoundly paradoxical one. The implosion of totalitarian state socialism (or its transformation to state capitalism as in the case of China and increasingly so in Russia) marked the end of two competing visions of what constitutes a "good" society. However, the historical victory of "democratic" capitalism effaced concern with democracy as the presumed equality of each and every one and hastened the transformation from a political to a managerial state on the one hand and, on the other hand, from democracy as the instituted space of agonistic encounter to the broadening and deepening of individual consumer choice and the hegemony of the market imperative as naturalized resource allocation technology within a consensually agreed socioeconomic order (Jörke 2005, 2008; Rancière 2005; Blühdorn 2006).

The Depoliticization of the Economy

The economization of politics parallels what Pierre Bourdieu diagnosed as "the depoliticization of the economic" (Bourdieu 2002). By this, he means the process through which the modalities by which natural and social resources are socially and economically appropriated, transformed, and distributed in the form of goods and services of all kinds no longer can be subjected to real public choice. Only the private ownership of nature and its organized transformation in a market-based configuration is deemed possible, desirable, and optimal. In such a naturalized market environment, any other modalities of organizing the production and distribution of life are deemed senseless. Public choice about the ownership of the commons of nature, the organization of production, and the distribution of the products is no longer possible or acceptable. Only relatively small alterations in the regulatory framework can be considered or debated—all this despite recurrent and deepening state interventions to assure that the private organization of capital circulation can be sustained.

Autocratic Governance

The combination of both these processes leads to the erosion of political control and accountability and, consequently, to the rise of more autocratic forms of governing that signal a reordering of the state-civil society nexus, whereby the state operates increasingly "at a distance" from the concerns, drives, and desires of large parts of civil society (Swyngedouw 2000). This is particularly evident in the emergence of new scalar and interscalar arrangements of governance (at both subnational and supranational scales such as urban development bodies, public-private partnerships, the European Union, the World Trade Organization, or G-20 meetings) that reorganize the institutional forms of governing as well as their scalar gestalt. As discussed in chapter 1, such arrangements of governance-beyond-the-state have become part of the system of governing, of organizing the "conduct of conduct."

Unauthorized Political Actors

The reordering of the gestalt of governance is accompanied by the extension of the regulatory and interventionist powers of authorities through the inclusion of what Ulrich Beck called "unauthorized actors" (experts, managers, consultants, and the like) in governance arrangements at all scales (Beck 1997). In addition, their proliferation is embedded in geographically variable configurations of a neoliberal polity that combines a desire to construct politically the market as the preferred social institution of resource mobilization and allocation, a critique of the "excess" of state, an engineering of the social in the direction of greater individualized responsibility, and the consolidation of the "tyranny of participation" (Cooke and Kothari 2001), often operating in what Maarten Hajer defines as an "institutional void" with opaque rules and procedures (Hajer 2003b).

Techno-Managerial Management

Agonistic debate[1] is increasingly being replaced by disputes over the mobilization of a series of new governmental technologies and modes of organization, managerial arrangements, and altered institutional forms that focus on accountancy rules and competitive performance benchmarking. "Doing politics" is reduced to a form of institutionalized social management, whereby problems are dealt with through enrolling managerial technologies and administrative procedures (Nancy 1992, 389). This public management of consensus relies on the opinion poll (rather than the ballot box), the perpetual canvassing of "popular" views, signaling the parameters of what needs to be governed and "policed." Post-democratic arrangements hinge, therefore, on the mobilization and normalization of a certain populism (Mudde 2004), something that is successfully mined by the rising popularity of right-winged populist movements and parties.

A Permanent State of Emergency

The nurturing of fear, its transformation into forms of crisis management, and the invocation of specters of pending catastrophe support these populist tactics (Swyngedouw 2007a). Consider, for example, the striking similarities in the mobilization of discourses of crisis by the elite expert-governance assemblage around questions like competitiveness, environment, immigration, terrorism, and the like (Badiou 2010; Baeten 2010; Swyngedouw 2010a). While economic crises and migration or environmental issues are a catastrophe for those who find themselves in the whirlwind of the negative effects they unleash, the elites do not tire to turn catastrophe for the excluded into a crisis-to-be-managed for the included. The catastrophe is always reserved for the excluded and powerless. Fear is a recurrent affective dispositive, but the techno-managerial elite insists on how proper interventions can be mobilized such that nothing really has to change, provided urgent and necessary institutional or technical measures are taken and implemented. We encounter here a politics that is sustained by the cultivation of a permanent "state of emergency," one that permits precisely pushing through certain visions and practices while disavowing, foreclosing, or silencing other possible avenues.

Žižek summarizes this emergent configuration under the banner of post-politics. The process of post-politicization is thus about the increasing administration (policing) of social, economic, ecological, or other issues, and this remains of course fully within the realm of the possible, of existing social relations: "The ultimate sign of post-politics in all Western countries," Slavoj Žižek maintains, "is the growth of a managerial approach to government: government is reconceived as a managerial function, deprived of its proper political dimension" (Žižek 2002a, 303). Politics becomes something one can do without making decisions that divide and separate (Thomson 2003).

Intensifying Outbursts of Antagonistic Violence

Post-democratization as consensus politics, however, inaugurates neither the disappearance of serial exclusion, radical socio-political conflict, antagonism, and occasionally violent encounter, nor greater political or socioeconomic inclusion. For example, the deterritorialization and denationalization of biopolitical relations, primarily as the result of growing diaspora nomadism, forced migration, and the explosion of multi-place networked identities, has given rise to truncated political rights, whereby some people are more equal than others in the exercise of political rights or commanding institutional powers that are still primarily territorial

(Swyngedouw and Swyngedouw 2009). Differential and unequal social and political citizenship rights, for example, are inscribed in or assigned to bodies depending on places of origin, destination, and patterns of mobility (Isin 2000). The geographically unequal and spatially fragmented political rights different individuals enjoy in different political-geographical settings—like the right to vote—are a case in point. Related to this, as Bob Jessop noted, "the scope of consensus politics is expanded to the whole of humanity but the presumed identity of the bare individual as *pars totalis* and a universal global humanity has been disturbed by a fundamentalism of identities that erupts onto the world stage" (Jessop 2005, 186). In other words, the universalizing procedures of consensus politics is cut through by all manner of fragmenting forces that often revolve around the resurgence of the "ethnic" evil, that is, identity politics as the object-cause that disrupts the consensually established order. While identitarian politics is loudly acclaimed, xenophobic or nationalist movements arise, whereby "incorrect" outsiders are violently excluded often through erecting all manner of new material, legal or other geographical barriers, walls, and camps (De Cauter 2004; Diken 2004; Minca 2005). In other words, post-democratic consensual procedures are cut through by all manner of often-disavowed antagonisms and recurrent violent outbreaks.

Indeed, consensus does not equal peace or absence of fundamental conflict. Post-politicization relies on both including all in a consensual pluralist order and excluding radically those who posit themselves outside the consensus. For the latter, as Giorgio Agamben argues, the law is suspended; they are literally put outside the law and treated as extremists and terrorists (Agamben 2005). Invoking the Whole/One of the people, while denying the antagonisms that cut through the social order, post-politicization is necessarily exclusive, partial, and predicated upon outlawing those who do not subscribe to the consensual arrangement. This is why for Agamben "the Camp" has become the core figure to identify the condition of our time. In other words, a Schmittian ultrapolitics that lurks behind and underneath the consensual order and does not tolerate an outside, that sutures the entire social space by the tyranny of the police (state) and squeezes out the political, pits those who participate in the instituted configurations of the consensual order against those who are placed outside, like undocumented workers, Islamists, radical environmentalists, communists, insurgents of a variety of kinds, commonists, alter-globalists, or the otherwise marginalized or politically inexistent. Those who rebel against the instituted order, whether it is the rioters in the French suburbs in 2005, the Spanish *Indignados*, the urban rebels in the UK in the summer of 2011, the Turkish insurgents on Taksim Square in 2013—all are invariably labeled by the political elites as scum, rabble, *racaille*, the *ochlos* who do not belong to the *demos*; they are the Rancièrian "part of no-part." They are often imagined and discursive, portrayed as prime drivers that undermine the cohesion or sustainability of the prevalent order.

What is at stake then, is the practice of real democracy, the public space for the encounter and negotiation of disagreement, for inaugurating a new sense and sensibility, where those who have no place, are not counted or named, acquire speech, or better still, appropriate voice, become visible and perceptible, and perform the egalitarian capacity to govern. While a consensual view refuses "to legitimize the centrality of antagonism in democratic politics, the post-democratic Zeitgeist forces the expression of this dissent through channels bound to fuel a spiral of increasingly uncontrolled violence [and] ... violent expressions of hatred which upon entering the depoliticized public sphere, can only be identified and opposed in moral or cultural (or eventually military) terms" (Stavrakakis 2006, 264–265). The rise of racism, violent urban eruptions, ethnic or religious rivalries, and more, become key arenas of social conflict (Žižek 2008c). In the absence of agonistic politicization of these antagonisms, they become expressed in outbursts of violence or, from a liberal cosmopolitan perspective, in the affective powers of humanitarian or ethical outrage (Kaika 2017, 2018).

In the absence of a recognized politicization of demands that are banned from the consensual order and that are not permitted to enter the public sphere of agonistic disagreement, violent encounter remains indeed one of the few courses open for the affective staging of active discontent. Of course, such manifestation of disagreement and dissent signal the possibility for a return, a re-treating, of the political. The post-democratic consensus and processes of depoliticization do not efface the political fully. Depoliticization is always incomplete; it leaves a trace and hence, the promise of a return of the political, a return, in Žižek's words, of the repressed. And it is this that we will turn to in chapter 3.

3 Theorizing the Political Difference: "Politics" and "the Political"

Political activity is whatever shifts a body from the place assigned to it or changes a place's destination. It makes visible what had no business being seen, and makes heard a discourse where once there was only place for noise. (Rancière 1998, 30)

Politics revolves ... around the properties of spaces and the possibilities of time. (Rancière 2006b, 13)

The Political Paradox

In this chapter, I shall first propose and explore an understanding of the political that foregrounds the presumption of equality as the axiomatic, yet contingent, foundation of democracy, that considers égaliberté (see Balibar 2010) as an unconditional democratic principle, and sees the political as a procedure that disrupts any given socio-spatial order by staging equality and exposing a "wrong." This "wrong" is a condition in which the axiomatic principle of equality is perverted through the institution of an order that is always necessarily oligarchic and at least partly unequal in its procedures of including actors and in the exercise of power. In the second section, I shall consider the contours and conditions of a return of the democratic political understood as a political order contingent on the presumption of equality of each and every one in their capacity to act politically. I shall argue that democracy and democratic politics, and the spaces for democratic engagement need to be taken back from the post-political and post-democratic oligarchic constituent police order that has occupied and filled out the spaces of instituted democracy.

One of the possible entries into the diagnosis and framing of post-democratic forms of governing is offered by a number of primarily, albeit not exclusively, French political philosophers and theorists. In lieu of considering democracy as an institutional regime characterized by a series of formal and informal norms and procedures, as a sphere of open pluralist communication and deliberation, and/or as a procedure for achieving negotiated consensus in the context of a plurality of positions, the argument developed here centers on the difference between "politics" on the one hand and "the political" on the other. Despite significant theoretical and political disputes between the various interlocutors in this debate, they share a concern with thinking through the difference between politics/polic(e)y (*la politique*) and the political (*le politique*). Paul Ricouer called this distinction the "Political Paradox" (Ricoeur 1965). The political refers to a broadly shared public space, a rational idea of living together, and signals the absence of a foundational or essential basis (in the social, the cultural, or political philosophy) on which to found a polity or a society. This is a quintessentially post-foundational position whereby the political is understood as a process of self-foundation on the basis of the absent ground of the social. In other words, the political expresses the nonexistence of society, stands for the incoherence and lack of foundation of society; it is the immanent field through which society as a more or less coherent order becomes instituted.

Politics, in contrast, refers to the power plays between political actors as well as the everyday choreographies of policy-making within a given institutional and procedural configuration in which individuals and social groups pursue their interests. In other words, politics refers to the empirically verifiable and institutionally articulated actions, strategies, and assemblages of governance that mark the management of the public sphere and is enacted through the interplay between various interests and positions. Politics institutes society, and provides coherence and a semblance of order. The "political difference" between the ontological foundation of the political as absent ground on the one hand and the everyday ontic practices of politics on the other is irreducible, yet there is a tendency for the latter to colonize, suture, and occasionally disavow the former. It is precisely this process of suspending the political that is at the heart of the analysis presented in this book. This is not an attempt to foreground the political as more important than actually existing instituted politics. The latter are of course central in the process of instituting any social order. The point is to consider the articulation between politics and the political, whereby the latter is a vital part in the process by which the former is transformed. The retreat of the political, I argue, fatally undermines the democratic substance of the former, and this is the central argument advanced here.

This ontological-ontic difference between *le politique* and *la politique* has been the leitmotiv of much of post-foundational political thought, of left Lacanians and of a number of radical political thinkers like Jean-Luc Nancy, Jacques Rancière, Claude Lefort, Chantal Mouffe, Pierre Rosanvallon, or Alain Badiou (see Hewlett 2007; Marchart 2007).[1] While there are significant differences between them (see, among others, Isin 2002; Dean 2006; Stavrakakis 2006; Marchart 2011), they share the view that the political marks the antagonistic differences that cut through the social, signaling the absence of a principle on which a society, a political community, or "a people" can be founded. In contrast to foundational perspectives that assume that politics is the way through which society realizes itself on the basis of a pregiven principle of what or who "the people" are or should be, post-foundational thought insists on the impossibility of the social, on the nonexistence of "the people" prior to its inauguration.[2] The "people" do not preexist the political sequence through which it is called into being as a procedure of living-in-common. It is this lack of foundation, the gaping hole (or void) in the social that renders founding the people impossible and that inaugurates the political as a marker of this gap or void. Politics, then, in the forms of the institutions and technologies of governing, and the tactics, strategies, and power relations related to conflict intermediation and the furthering of particular partisan interests, contingently institute society, and give society some (instable) form, and temporal and spatialized coherence (in a democratic polity, it does so in the name of "the people"). It is through politics that society comes into being in the form of a more or less functioning whole, a socio-spatial distribution and allocation of people, things, and activities, quilted through forms of institutionalization, modes of representation of the social order, and routinized or ritualized practices of encountering, relating, and exercising power. In a liberal democratic polity, the object of governing is the biopolitical "happiness" of the population. But it is also this procedure that sutures or colonizes the space of the political, and through this, forecloses the political origins of politics. Politicization as the procedure of instituting forms of governing the people is always accompanied by specific forms of depoliticization. Post-politicization is one such form, alongside others that Rancière identifies as parapolitics, metapolitics, or archipolitics (Rancière 1998; Bosteels 2014; Van Puymbroeck and Oosterlynck 2014).

The field of politics/the political is thus split into two: on the one hand the political stands for the constitutive lack of ground, while on the other hand politics stands for the ontic, the always contingent, precarious, and incomplete attempt to institutionalize, to spatialize the social, to offer closure, to suture the social field, to let society coincide with "community" understood as a cohesive and inclusive whole. Therefore, this ontic level always harbors totalitarian moments, exclusive operations, various forms of marginalization, silences, and inequities; politics implies the "retreat of the political." This retreat is marked by a colonization or disavowal of the political by politics or the sublimation of the political by replacing it with community (as an imagined coherent unity), a particular sociological imaginary of "the people" (as nation, ethnic group, religious community, or other social category or form of social bonding), "organization," "management," "good governance." As Olivier Marchart attests, the colonization of the political by politics operates through "the merging of the political with diverse authoritative discourses, among which [Lacoue-Labarthe and Nancy] count socio-economic, technological, cultural, or psychological discourses. In the process, the political converts itself ... into "technological" forms of management or organization; a process which leads to the effective silencing of genuinely political questions" (Marchart 2007, 66). Politics is reduced to institutionalized social management, whereby recognized social and other problems are dealt with through administrative-organizational-technical means and questioning things as such disappears (Nancy 1992, 389). Post-democratization, as outlined in chapter 2, is a particular procedure of colonization of the political predicated upon the disavowal or foreclosure of antagonism through the progressive inauguration and institutional arrangement of consensus in a pluralistic liberal order and the depoliticization of the sphere of the "economic" understood as the procedures of wealth creation and distribution.

Similar conclusions, albeit from rather different theoretical positions, are proposed by what Yannis Stavrakakis calls the Lacanian Left. For Lacanians (see Stavrakakis 2006), politics always operates in and through a symbolic order, the quilting of a chain of signifiers that assigns everyone and everything a certain place and function within the social edifice.[3] In other words, the symbolic order (or the Law) in which we dwell and through which we understand our (and everyone else's) place in the world, institutes society (Castoriadis 1987). Politics (as an instituted symbolic order) invokes and constructs a mode of living together, a whole, that becomes (re-)presented in the interplay between political forces, programs, parties, civil society organizations, formal and informal routines of living everyday life, and the like. There are a variety of possible ways through which the community as living-in-common becomes instituted through the symbolic framing of politics. Consider, for example, how ethnic, religious, national, or other forms of collective fantasies become the imaginary ground for instituting society politically (Kaika 2011), a process that simultaneously represses or disavows the antagonisms (such as class divisions) that cut through any social formation. Such symbolic formations fail inherently to suture fully the social order. A gap, a void, a lack, or excess always remains and resists symbolization, a hard kernel that is not accounted for in the symbolic order. This remainder or surplus (what Lacanians call the "Real")—in other words, that which cannot be symbolized by the existing interplay of political forces but whose uncanny presence is invariably disturbingly sensed—disrupts and destabilizes. This hard kernel, the bone of the unsymbolized Real that haunts and occasionally resurfaces in symptomatic ways stands as guarantee for the return of the political. The political—as the return of the repressed—is the affirmation and affective appearance of the unsymbolized in the order of the police/politics and signals or expresses a disruption in the order of the sensible, the transformation of the aesthetic register. This aims at the institution of new "radical" imaginaries (Castoriadis 1987; Kaika 2010). The political, therefore, revolves around the Real of the situation, the hard bone that resists symbolization, but that nevertheless reappears in the form of a performative act that cuts through the existing order. It is excessive to the situation, interrupts the institutional frame, and in doing so, attests precisely to what is foreclosed, disavowed, or repressed by the everyday practices of existing rules, routines, and procedures. This Lacanian perspective also holds that the political signals the empty ground of the social, the inherently split condition of being that prevents a being-together to emerge outside a properly constituted symbolic order. However, the symbolic order is always incomplete; there is a stubborn remainder that resists symbolization. For example, when a democratic form became institutionalized in France after the French revolution on the basis of the principle of universal equality among equal citizens, this

symbolization disavowed the existence of the majority of noncitizens, who
were devoid of any political right. They would, in the nineteenth century,
constitute themselves as the proletarian political subject, a *Demos*, whose
struggle and process of subjectivation unfolded precisely around intruding
this symbolically constituted order to which they were the unnamed and
nonsymbolized remainder or excess. They claimed the right to be part of
the order; they demanded the right to have rights as declared in the constitution, yet denied in the everyday reality of politics. Those un(ac)counted
and nonexistent in the instituted order became the stand-in for the universality of "the People" as *Demos*. The American Civil Rights movement
similarly unleashed a process of staging equality and demanding egalitarian
inclusion in the constituted order. Are today's undocumented immigrants,
claiming inclusion, not a contemporary example of the political paradox,
in other words, the promise of equality that is disavowed in the policing,
coding, categorization, enclosing, and naming of some as outside the symbolic order of the Law? Are they not precisely the excessive component that
expresses the lie at the heart of the democratic configuration—that some
are more equal than others? The foreclosure of the Real of class antagonism
in post-democratic discourse equally suggests the disavowal of key antagonisms that cut through the social.

If this hard kernel is disavowed or foreclosed, as in the present consensual post-democratic configuration, this "return of the Real" is displaced, in
a short-circuiting of the antagonisms that mark the political, from the terrain of the political to the domain of moral outrage or compassion and the
ethical concern for the subaltern "Other" (Kaika 2017) on the one hand, or
to the ultrapolitical domain of unmediated repression, xenophobic nation-
alism, exclusion, and violence on the other (see Žižek 2008b, 2017). A case
in point is the unrelenting "clash of cultures" between "Western" values
and Islamo-fascist interruptions and their nurturing on both sides of the
identitarian divide.

The preceding theorization begs the question of what constitutes the
democratic political. For Claude Lefort, the foundational but contingent
inaugural political gesture of democracy is the insistence on the "empty
place of power." In contrast to a divine or imperial order, anyone can temporarily claim the place of power in a democratic constellation (Lefort
1989, 1994). This dovetails with the presumption of equality—as axiomatically given and presupposed and not as a sociologically verifiable

concept—upon which the democratic invention is contingently grounded. This presumption stands in strict opposition to the sociologically verifiable inequalities and differences (along gender, class, or other recognized classifications) that characterize any given order, including instituted democratic politics. This "emptiness" not only refers to the condition in which the place of power is not transcendentally given and instituted (by God or in the person of the king/emperor), but is fundamentally undecided precisely because of the contingent presumption of equality of each and all qua political beings, or, in other words, because of the equal capacity of each to act politically. Anyone can claim the place of power in democracy: there is no symbolic authority (in the social, natural, or divine), no Big Other, that designates the site of power; there is nothing in the social but the recognition of its own absent ground, of the self-reflexive understanding that the social is split and, therefore, radically heterogeneous. For Lefort, democracy as the contingent affirmation of society's nonexistence (as a coherent goal or assemblage) requires institutionalization of the social order on the basis of antagonism and dissensus. Democracy is "founded upon the legitimacy of a debate as to what is legitimate and what is illegitimate—a debate which is necessarily without any guarantor and without any end" (Lefort 1989, 39).

The truth (in the sense of being true or faithful to something) of democracy lies in the axiomatic presumption of equality and the free expression of its egalitarian practices. Etienne Balibar (1993, 2010) names this fusion of equality and freedom *égaliberté*, the former defined as the absence of discrimination and the latter as absence of repression (Dikeç 2001). This very promise of the democratic political, which is scandalously perverted in its (post-democratic) institutionalized form and necessitates continuous reclamation, inaugurates the universalizing and collective process of emancipation as *égaliberté*. Indeed, freedom and equality can only be conquered and occupied; they are never offered, granted, or distributed. For Chantal Mouffe, there is a paradox between the recurrent need to produce some sort of order through institutional arrangements (like the state) on the one hand and the contingent, inherently questionable, never fully closed, and antagonistic givens of the social on the other hand (Mouffe 2000). What the democratic invention asserts, however, is the translation of antagonistic hatred into agonistic adversaries, the axiomatic invocation of equality, the possibility of producing arrangements that simultaneously acknowledge the inherent antagonisms in the social and permit their expression in institutional or other organized forms.

Staging Egalitarian Spaces: The Political as Immanence

In this section, I shall explore further the notion of the political that foregrounds equality as the contingent presumption upon which democracy rests and its implications for thinking through the articulation between the political and politics today. While symptomatically appropriating elements from the preceding perspectives on the political difference, and despite significant differences, post-Althusserian thinkers like Jacques Rancière, Alain Badiou, Etienne Balibar, and Slavoj Žižek reject the attempt to refound political philosophy on the basis of the "political difference" as articulated earlier. Their conceptualization is driven by a fidelity to the possibility of a revived emancipatory political project that moves beyond the efforts to reinscribe political democracy in philosophical thought (Bosteels 2005; Badiou and Žižek 2010). Theirs is a view that understands the political as a retroactively revealed moment of eruption, an event, a political act that contingently might open a procedure of disrupting a given socio-spatial order, one that addresses a "wrong"[4] in the name of an axiomatic and presumed equality of each and every one (Bassett 2008, 2016). This "wrong" is a condition in which the presumption of equality is perverted through the contingent socio-spatial institution of an oligarchic and empirically verifiable police order. The political arises then in the act of performatively staging equality, a procedure that simultaneously makes visible the "wrong" of the given situation and demonstrates equality. For example, when, on December 1, 1955, Rosa Parks sat down on the "wrong" seat on the bus in Montgomery, Alabama, she simultaneously staged equality and exposed the inegalitarian practice of a racialized instituted order, despite the latter's constitutional presentation of equality (May 2008). Although she was of course not the first activist to engage in this defiant act, her act became retroactively constituted as one of the founding events of the political sequence that today goes under the name of the Civil Rights movement. Indeed, a political sequence unfolds through the universalization of such particular place-moments and localized acts, when others declare "fidelity," to use Alain Badiou's terminology, to the inaugural egalitarian event and, in doing so, publicize and universalize the egalitarian and emancipatory act it stages. It is such a political sequence that may begin to affect and occasionally succeeds in transforming the instituted order. The Civil Rights movement in the United States was exactly such a procedure of emancipatory universalization of a geographically specific interruption of an inegalitarian practice. It is the moment when a concrete and particular socio-spatial condition (racial segregation in public transport, in this case) becomes the stand-in for a generalized democratic and egalitarian demand ("we are all Rosa Parks") that cuts through generalized institutionalized racism, a particular that stands in for the whole of the people and becomes recognized as such.

The theorizations of such re-treating or reemergence of the political vary as different interlocutors (Rancière, Badiou, Žižek, or Mouffe) locate the modalities of its immanent appearance in different registers (Wilson and Swyngedouw 2015). For example, while Rancière maintains a radical disjuncture between the dissensual spatiality of democratic politics and the consensual order of the police and, thus, an agonistic position with respect to the possible institutionalization, the ultimate idyll of real democracy (for a detailed argument, see May 2008, 2010), Badiou and Žižek insist that the democratic and emancipatory sequence operates under the generic name of "communism" (see Badiou 2008a; Žižek 2008b) as a practical and immediately realizable possibility (see also Bosteels 2011). My argument in this chapter, however, concentrates on the similarities, which revolve around the notions of emancipatory and egalitarian politics, a critique of the post-political order, the presumption of equality, and the democratic political as asserting/staging equality within a geographically specific and diverse inegalitarian frame. We shall take up the question of communism again in the final chapter.

Let me begin with considering Jacques Rancière's conceptualization of politics and the political. He explores whether the political can still be thought of in a context in which a post-democratizing consensual policy arrangement has increasingly reduced politics to policing, to managerial consensual governing (Rancière 1995b, 1998). Rancière distinguishes conceptually "the police/policy" (*la police*), "the political" (*le politique*), and "politics" (*la politique*). He understands the political as the place of encounter, the meeting ground of two heterogeneous processes: the process of governance, or the police, on the one hand and the process of emancipation, or politics, on the other.

The "police" is defined as the existing order of things and constitutes a certain "partition of the sensible" (Rancière 2001, 8): the police refers to "all the activities which create order by distributing places, names, functions" (Rancière 1996, 173). Here, Rancière is on terrain similar to but expanding on Michel Foucault's thesis. For Foucault, the police refers to a governmental dispositive emerging in the seventeenth century that signals a shift in state power from sovereign to bio-power. The police refers to all those activities, interventions, and regulations that nurture and sustain life, that permit life to be lived, but in the process also decides who lives and who dies, what lives are worth living or not. While the police includes the formal institution of the police as repressive force, it encompasses also all state and nonstate activities that impinge on the continuation of life (the allocation of things and people in social and physical space, food, health, education, administration, social welfare for the deserving and the like (for details, see Foucault 2004a, 2004b; Elden 2007; or Dikeç 2005) and organizes a distribution of the perceptible. The police comprises simultaneously a set of formal and informal rules and procedures that are shared by the members of a polity in specific places and around specific activities (Lévy, Rennes, and Zerbib 2007) while also partitioning (distributing) them. This partitioning of the sensible "renders visible who can be part of the common in function of what he does, of the times and the space in which this activity is exercised. … This defines the fact of being visible or not in a common space. … It is a partitioning of times and spaces, of the visible and the invisible, of voice and noise that defines both the place (location) and the arena of the political as a form of experience" (Rancière 2000b, 13–14).

The police refers to both the activities of the state as well as to the ordering of social relations and "sees that those bodies are assigned by name to a particular place and task; it is an order of the visible and the sayable that sees that a particular activity is visible and another is not, that this speech is understood as discourse and another as noise" (Rancière 1998, 29). To return to the example of Rosa Parks, she would have remained invisible if she had taken a "proper" seat on the bus, the place allocated to non-whites. By explicitly refusing to do so and theatrically enacting equality, her act became visible; it cut through the "partition of the sensible." The police classifies, orders, distributes. As Dikeç maintains, the police "relies on a symbolically constituted organization of social space, an organization that becomes the basis of and for governance" (Dikeç 2007, 19). The basic gesture of the police, therefore, is not power. It is distribution: the allocation of things and people, the temporal and spatial organization of activities, assigning functions and their interaction: "The police are a rule governing the appearance of bodies that configures a set of activities and occupations and arranges the characteristics of the spaces where these activities are organized or distributed" (Rancière 1998, 29).

As such, the police is an aesthetic affair that renders and sustains certain things, places, utterances, and conditions as common sense, while others remain in the register of non-sense. The "sensible" refers both to what is acceptable and naturalized as common sense as well as to an "aesthetic" register as that which is seen, heard, and spoken, which is registered and recognized. Planning, urban policies, health and labor policies, environmental rules and practices, and the like are of course key dispositives of the police.

Against the Althusserian notion of the police as a form of interpellation, a process of "hailing" by the Repressive State Apparatuses, Rancière focuses on circulation and distribution. For example, when a police officer calls "Hey you there—" the affective force is one of ideological interpellation sustained by state power. You feel spoken to and summoned to behave and/or act in a certain manner. In contrast to this, Rancière's notion of the police is not so much articulated around power and hailing, but rather more on the importance of doing what one is supposed to within the socio-spatial order. For example, when a police officer tells people, "Move on, there is nothing to see here" (when something transgressive happens in public space), the affective force is one of assigning and distributing "proper" places, functions, and activities, to assign some as legitimate and others as non-sense and, thus, illigitimate.

Forms of excess are either incorporated in the police's symbolic order ("keep on shopping") or ruthlessly repressed. The police confers the Platonic ideal of a fully ordered society in which everyone takes their places and performs the functions allocated to each place, while expelling forms of acting and places of being that are surplus or excess to the given situation. The police order is predicated upon saturation, upon suturing social space without outside, excess, or remainder—it aspires to include all and everyone: "The essence of the police is the principle of saturation; it is a mode of the partition of the sensible that recognizes neither lack nor supplement. As conceived by 'the police,' society is a totality compromised of groups performing specific functions and occupying determined spaces" (Rancière 2000a, 124) and in which all are included and assigned a proper place.

The political, in contrast, manifests itself precisely in forms of excessive or supernumerary acting. Indeed, the drive to suturing is of course never fully realized as other modes of being-in-common or forms of communing, of constructing a political community are constantly emerging, yet are not recognized as such. This manifests in what we discussed earlier as the return of the repressed. They are enacted by the "part of no-part," those bodies and activities that are not recognized as part of the social totality. Indeed, the constitutive antagonisms that rupture society preempt saturation; there will always be a constitutive lack or surplus; that which is not accounted for in the symbolic order of the police (Dikeç 2005), which is only registered as noise. It is exactly this lack or excess that keeps the space of the political vacant or empty, and guarantees "its return." Such moments can be understood as material and symbolic time-place events that shatter the oligarchic distribution of the sensible, disrupt the perceptible, and recast the coordinates of what constitutes the social by exposing the ruptures that cut through the social body. The "subjective" form of violence they embody reveals the "objective" violence sustained by a police order that disavows the "wrong" that is exposed through the intervention (Žižek 2008c). The spatialization of events like these signal the affirmation of the political and is about enunciating dissent and rupture, literally voicing speech that claims a place in the order of things, demanding "the part of those who have no-part" (Rancière 2001, 6). This immanence of *Demos*, those without a part staging their equality, becomes the kernel that stands in for the whole of the people. Evental moments like these are invariably seen as insurgencies in the eyes of the police and take specific geographical forms.

The political, thus, is about enunciating dissent and rupture, literally voicing speech that claims a place in the order of things. It is the arena in which *Ochlos* (rabble, mob, multitude) is turned *Demos* (People), where the anarchic noise of the rabble (the part of no-part as Rancière would name it) is turned into the recognized voice of the people, the spaces where what is only registered as noise by the police is turned into voice. The political is, therefore, always disruptive; it emerges with the "refusal to observe the 'place' allocated to people and things (or, at least, to particular people and things)" (Robson 2005, 5): it is the terrain where the axiomatic principle of equality is tested in the face of a wrong. Such wrong is the experience

and practice of inequality that inheres in the oligarchic spaces of an instituted polity. In other words, the axiomatic albeit utterly contingent foundational gesture of democracy is equality. The political opens up a lived
space through the testing of a wrong that subverts inequality by staging
and performing equality. Despite many disagreements, here Rancière is on
a terrain explored by Alain Badiou as well: "Equality is not something to be
researched or verified but a principle to be upheld" (Hallward 2003a, 228).
For Badiou, the Rancièrian anarchic interruption is nonetheless insufficient
without fidelity to the inaugural event—to the moment of interruption—
a fidelity that ushers in forms of organization, militant activism, and the
experimental work of testing "the Communist Hypothesis" (see also chapter 8). An emancipatory political sequence, Badiou argues, unfolds through
militant political subjects' universalizing fidelity to the presumption of
equality staged in the inaugural event. It is through this faithful acting that
the truth of the political sequence asserts itself.

In *Nights of Labour*, for example, Rancière explores how workers in
nineteenth-century France, through carving out, taking, and producing/
staging their time and space, became political subjects under the name of
the proletarian (Rancière 1989). He documents how workers during the
time that was their own (in the night, that is, after or before work) choreographed their own spaces of writing, imagining new social systems and
modes of being, through talking, singing, composing, discussing—in short,
through the practice of imagining a different equality. Emancipation was,
therefore, about taking the right to their own time and their own place,
to produce their own geographies and histories, to think, to play, to seize
the terrain that was allocated to the bourgeoisie. Emancipatory politics is
the refusal to be restricted to the places distributed by the police order (the
factory/the home/the bar/the shopping mall); it disrupts and declassifies,
and claims what is not authorized. This example also illustrates the radical
departure from other theories and perspectives of emancipatory change.
There is neither privileged sociological nor philosophical theorization that
calls a political subject into being (like class for Marx, women for feminists, and so on), nor a-priori produced political subjects (like "proletar-
ians" or "feminists"). Political subjectivation unfolds in and through the
staging/enacting of equality that exposes a "wrong" in the inegalitarian
distribution of the sensible. In doing so, the subject appears in space and
transforms both him-/herself and the socio-spatial configuration through

performative practices of dissensual spatialization. This process involves, as Davis (2010, 84) points out, "argumentative demonstration, a theatrical dramatization and a 'heterologic' disidentification." This is the process of becoming political subject, a subjectivation that is "never simply the assertion of an identity but the refusal of an identity imposed by others, by the police order" (ibid, 88). By demonstrating equality as theatrical publicity, subjectivation implies a process of insurrectional presentation of egalitarian difference, one that renders visible and sensible the censored (by the police) practices of equality and freedom. It is this meeting point of police and politics that Rancière defines as "the political." The political occurs as the frictional confrontation between the specific socio-spatial configuration on the one hand and the universalizing gesture of equality on the other. Peter Hallward summarizes:

> If the supervision of places and functions is defined as the "police," a proper *political* sequence begins, then, when this supervision is interrupted so as to allow a properly anarchic disruption of function and place, a sweeping de-classification of speech. The democratic voice is the voice of those who reject the prevailing social distribution of roles, who refuse the way a society shares out power and authority. (Hallward 2003b, 192)

The political arises when the given order of things is questioned; when those whose voice is only recognized as noise by the police order claim their right to speak, acquire speech, and produce the spatiality that permits exercising this right. Such declassification disturbs all representations of places, positions, and their properties (Rancière 1998, 99–100) and destabilizes socio-spatial orderings and the principles of their distribution.

This is an inherently public affair and unfolds in and through the transformation of space, both materially and symbolically, redefines what constitutes public or private space and its boundaries, and re-choreographs socio-spatial relations. Such view of the political as a dissensual space stands in sharp contrast to the consolidating consensual "post-politics" of contemporary neoliberal "good" governance and its institutionalized post-democratic counterpart.

Equality is, consequently, not some sort of utopian longing (as in much of critical social theory) or a sociological and empirically verifiable category, but the very axiomatic condition upon which the democratic political is founded and that becomes symbolized, named, and sensible in the actual process of its staged enactment. The very promise of the democratic, but which is always scandalously perverted in instituted forms of governance, and therefore necessitates its continuing reclamation, is founded on the universalizing and collective process of emancipation as *égaliberté*. The political, therefore, is not about expressing demands to the elites to rectify inequalities or unfreedoms, or a call on the state exemplified by the demands of many activists and others who are choreographing resistance to the police order, but, in contrast, it is the demand by those who do not count to be counted, named, and recognized. It is the articulation of voice that stages its place in the spaces of the police order: it appears, for example, when undocumented workers shout, "We are here, therefore we are from here," and demand their place within the socio-political edifice or when the Spanish *Indignados* demand "Democracia Real Ya!" or the Occupy movement's claim to be the 99 percent who have no voice. These are the evental time-spaces from where a political sequence may unfold (see chapter 7).

The political emerges through the act of political subjectivation, the embodied decision to act, to interrupt, to stage; it appears as the voice of "floating subjects that deregulate all representations of places and portions" (Rancière 1998, 99–100) and that occupies, organizes, and restructures space: "In the end everything in politics turns on the distribution of spaces. What are these places? How do they function? Why are they there? Who can occupy them? For me, political action always acts upon the social as the litigious distribution of places and roles. It is always a matter of knowing who is qualified to say what a particular place is and what is done to it" (Rancière 2003b, 201).

The political, therefore, always operates at a certain minimal distance from the state/the police and customarily meets with the violence inscribed in the functioning of the police (Abensour 2004). Its spatial markers are not the parliament, meeting room, or council chamber, but the public square, the housing estate, the people's assembly, the university campus, the social center, the factory floor.

The possibility of political change immanently emerges as the moment or process of confrontation with the police order, when the principle of equality confronts a wrong instituted through the police order. It appears thus when the police order is dislocated, transgressed, "when the natural order of domination is interrupted by the institution of a part of those who have no part" (Rancière 1998, 11). The space of the political is to "disturb this arrangement [the police] by supplementing it with a part of the no-part identified with the community as a whole" (Rancière 2001, Thesis 7); it is a particular that stands for the whole of the community and aspires toward universalization. And of course, such transformation unfolds through the production of concrete spaces and geographies, and the recognition of the principle of dissensus. As Rancière attests: "The principal function of politics is the configuration of its proper space. It is to disclose the world of its subjects and its operations. The essence of politics is the manifestation of dissensus, as the presence of two worlds in one" (Rancière 2001, Thesis 8); it occurs when there is a place and a way for the meeting of the police process with the process of equality (Rancière 1998, 30), when a new world and new way of "worlding," of making a new world, becomes present in the world, when two worlds meet and make their presence felt. Politics understood in these terms rejects a naturalization of the political, signals that a political act does not rely on expert knowledge and administration (the partition of the sensible), but on a disruption of the field of vision and of the distribution of functions and spaces on the basis of the principle of equality. This view of politics as a space of dissensus, for enunciating difference and for negotiating conflict, for experimenting with a new (common) sense and forms of sensuous being, stands in sharp contrast to the consolidating consensual "post-politicizing" rituals of contemporary neoliberal governance. Of course, this argument also begs the question as to what to do, how to reclaim the political, as discussed in this chapter, from the debris of consensual autocratic post-democracy?

Challenging Post-Democratization: From Staging Equality to Producing Egalitarian Spaces

As noted, much of the geographical literature on democratic spaces, empowerment, and the public sphere articulates around processes of privatizing/enclosing and re-policing material or symbolic public space under neoliberalization, the practices and configurations of "public" spaces, and the modalities of nonoppressive action, the injunction to be hospitable, and the need for inclusive ethical encounter. However important these perspectives are, the theorizations explored here invoke a different articulation between space and the political/politics. They focus on how egalitarian political spaces can be claimed, staged, choreographed, and materialized in both the physical and symbolic sense (Merrifield 2011). They unfold around the central tropes of emergence, insurrection, equality, and theatrical staging through performative acting, and foreground the political act as a located intervention/interruption and the procedures of spatializing equality (Springer 2010). For Slavoj Žižek, politics occurs

with the emergence of a group which, although without a fixed place in the social edifice (or, at best, occupying a subordinate place), demanded to be included in the public sphere, to be heard on an equal footing with the ruling oligarchy or aristocracy, i.e. recognized as a partner in political dialogue and the exercise of power. ... Political struggle proper is therefore not a rational debate between multiple interests, but the struggle for one's voice to be recognized as the voice of a legitimate partner. ... Furthermore, in protesting the wrong they suffered, they also presented themselves as the immediate embodiment of society as such, as the stand-in for the Whole of Society in its universality. ... Politics proper thus always involves a kind of short-circuit between the Universal and the Particular: the paradox of a singular which appears as a stand-in for the Universal, destabilizing the "natural" functional order of relations in the social body. (Žižek 2006b, 69–70).

The emergence of politicization is always specific, concrete, and particular, but stands as the metaphorical condensation of the universal. This procedure implies the production of new material and discursive spatialities within and through the existing spatialities of the police. It asserts dissensus as the base for politics. Politics has, therefore, no foundational place or location, it cannot be theoretically posited or socio-spatially located a priori. It emerges; it is an emergent property. It can arise anywhere and everywhere: "Space becomes political in that it ... becomes an integral element of the interruption of the 'natural' (or, better yet, naturalized) order of domination through the constitution of a place of encounter by those that have no part in that order. The political is signaled by this encounter as a moment of interruption, and not by the mere presence of power relations and competing interests" (Dikeç 2005, 172). An important rupture opens here between Rancière on the one hand and Badiou on the other. Rancière insists on the impossibility of the institutionalization of democracy and, consequently, on the abyss between any instituted order (the police) and the democratic presumption of equality (politics):

The community of equals can always be realized, but only on two conditions. First, it is not a goal to be reached but a supposition to be posited from the outset and endlessly reposited. ... The second condition ... may be expressed as follows: the community of equals can never achieve substantial form as a social institution ... A community of equals is an insubstantial community of individuals engaged in the ongoing creation of equality. Anything else paraded under this banner is either a trick, a school, or a military unit. (Rancière 1995a, 84)

Institutionalization is therefore always necessarily incomplete; the terrains of the political and of the police remain forever lodged in different registers. Insurgent activism and the staging of egalitarian practices, the principled anarchism of insurgent architects of emancipatory struggles, exhaust the horizon of egalibertarian politics. Badiou does not share this view of the emancipatory political as merely anarchic disruption. A political truth procedure, for Badiou, is initiated when in the name of equality a militant fidelity to an event is declared, a fidelity that, although always particular, aspires to become public, to universalize. It is a wager on the truth of the egalitarian political sequence and the possibility of its realization (Badiou 2005a), a truth that can be only verified ex post. Preferred examples of Badiou and Žižek are Robespierre, Lenin, or Mao in their declaration of fidelity to the procedure of communist truth in the revolutionary event. For Badiou, communism is an idea related to a generic human destiny and articulates with the hypothesis that another collective organization (than the present inegalitarian one) is practicable and immediately realizable (Badiou 2010). If a political (communist) truth sequence unfolds through the declaration of fidelity to an inaugural event, one that can be only retroactively discerned as such, it is the unfolding of the theoretical and practical production of this truth procedure that will decide its historical-geographical realization. The latter calls for forms of political organization that operate "at a distance from the State," not as part of the "state of the situation" (or police order).

Irrespective of these theoretical differences, democratic political spaces are active moments of construction of new egalitarian spatialities inside and through the existing (public and private) geographies of the police order. Such production operates through the (re-)appropriation of space, the production of new spatial qualities and new spatial relations, both materially and symbolically, and expresses what Castoriadis would call a radical imaginary at work. The aforementioned example of Rosa Parks and the Civil Rights movement is a case in point. Another example of such political sequence erupted when, in 1981, *Solidarnosc*'s demands for better working conditions on the Gdansk shipyards translated into the universal demands for political rights against the oligarchic bureaucratic order of the socialist form of state capitalism and their apparatchiks in Poland; when the latter acknowledged the demands of the activists, their police order's symbolic edifice and constituted order crumbled and revealed the empty locus of power. The insurgents launched a proper political sequence that would overturn the symbolic order and the distribution of functions and places associated with it. Their subsequent history, of course, also signaled how the fidelity to equality crumbled through the colonizing process of incorporation into a post-political European police order.

It is within the aporia between Rancière and Badiou that today's insurgent spatial practices can be situated. If the political is increasingly foreclosed as post-democratization unfolds, reemerging sporadically as outbursts of irrational violence and insurgency as seemingly the only conduit through which to stage dissensus, what is to be done? What design for the egalitarian staging of political space can be thought? How and in what ways can the courage of the collective intellect(ual) be mobilized to think through dissensual or polemical spaces? I would situate the tentative answers to these questions in a number of interrelated registers of thought. We shall also return to this question in the final chapter of the book.

First, rather than embracing the multitude of singularities and the plurality of possible modes of becoming, rather than reveling in the immanence of imperial transformation, an immanence to which there is no outside (Hardt and Negri 2001), rather than the micropolitics of dispersed resistances, alternative practices, and affects (see Holloway 2002 or Critchley 2007), the view explored in this book foregrounds division and exclusion and emphasizes the "the political act" and a fidelity to a political truth procedure that necessitates taking sides (see Dean 2006, 115). The insurgent egalitarian performances that Rancière and others call for require, if they are to be effective, transgressing the fantasy of the sort of acting that sustains the post-political order and that calls upon "resistance" as a positive injunction. The act of resistance ("I have to resist the process of, say, neoliberalization, globalization, or capitalism, or otherwise the city, the world, the environment, the poor will suffer") just answers the call of power in its post-democratic guise. Resistant acting is actually what is invited, but leaves the police order intact. Politics understood merely as rituals of resistance is, according to Žižek, doomed to fail politically: "Radical political practice itself is conceived as an unending process which can destabilize, displace, and so on, the power structure, without ever being able to undermine it effectively—the ultimate goal of radical politics is ultimately to displace the limit of social exclusions, empowering the excluded agents (sexual and ethnic minorities) by creating marginal spaces in which they can articulate and question their identity" (Žižek 2002b, 101).

Such impotent political activism (the hysterical "passage to the act") that is merely aimed at the state and demands inclusion in the institutional registers does nothing but keep the state of the situation intact and contribute to solidifying the post-political consensus. Resistance as the ultimate horizon of social and political movements (see Featherstone 2008) has become a subterfuge that masks what is truly at stake—how to make sure that nothing really changes. The choreographing of resistance is not any longer concerned with transgressing the boundaries of the possible, acceptable, and representable, but rather a symptom of the deepening closure or suturing of the space of the political. The problem with such tactics is not only that they leave the symbolic order intact and at best "tickle" the police, they also are an active part of the process of post-democratization. As Žižek puts it, "these practices of performative reconfiguration/displacement ultimately support what they intend to subvert, since the very field of such 'transgressions' are already taken into account, even engendered by the hegemomic form" (Žižek 1999, 264).

The temptation to act within and upon the police as rituals of resistance has to be looked at with suspicion. However, the return of the political today in the form of violent urban outbursts and that I shall return to in chapter 7, outbursts that seem to be without vision, project, dream or desire, without proper symbolization, is also suspect. It is nothing but the flipside of the disavowal of the violence of consensual governance. They are the Manichean counterpunch to rituals of resistance. The return of the repressed or of the Real of the political in the form of urban violent insurgencies, redoubles in the violent encounter that ensues from the police order, whereby the rallying protesters are placed, both literally and symbolically, outside the consensual order, are countered and corralled by the violence of the police; they are nothing but, in the words of the former President of France, Nicolas Sarkozy, "scum" (*racaille*), people without proper place within the order of the given.

Second, attention needs to turn to the modalities of repoliticization. Repoliticizing space as an intervention in the state of the situation that transforms and transgresses the symbolic orders of the existing condition marks a shift from the old to a new situation, one that cannot any longer be thought of in terms of the old symbolic framings. For Žižek, such a political act does not start "from the art of the possible, but from the art of the impossible" (Žižek 1999). Emancipatory politics is thus about inaugurating practices that lie beyond the symbolic order of the police; about formulating demands that cannot be symbolized within the frame of reference of the police and, therefore, would necessitate transformation in and of the police to permit symbolization to occur. Yet, these are demands and claims that are eminently sensible and feasible when the frame of the symbolic order is shifted, when the parallax gap between what is (the constituted symbolic order of the police) and what can be (the reconstituted symbolic order made possible through a shift in vantage points, one that starts from the partisan universalizing principle of equality) is asserted. This is the democratic political process through which equality is tested and given substance, and that requires the transformation of socio-physical space and the institution of a radically different partition of the sensible. The form of politicization predicated upon universalizing egalibertarian demands cuts directly through the radical politics that characterize many contemporary forms of resistance. This could be glimpsed in the democratizing outbursts in the streets of Tunis, Athens, or Madrid in 2011, or Istanbul in 2013. It is also the sort of demand expressed when undocumented and other immigrants in Europe or the United States claim their egalibertarian place.

Third, the proper response to the injunction to undertake action, to design the new, to be different (but that is already fully accounted for within the state of the situation), is to follow Bartleby's modest, yet radically transgressive, reply to his Master: "I would prefer not to ...": the refusal to act, to stop asking what they want from me, to stop wanting to be liked. The refusal to act is also an invitation to think or, rather, to think again; to try to symbolize the desire expressed in the act of refusal to obey the injunctions of the Master discourse. There is indeed an urgent task that requires the formation of new egalitarian imaginaries/fantasies and the resurrection of thought that has been censored, scripted out, suspended. In other words, the key question to be posed today is: Is it still possible to *think* the design of democratic, polemical, equitable, free common spaces for the twenty-first century? Can we still think through the censored metaphors of equality, communism, living-in-common, solidarity, and proper political democracy? Are we condemned to rely on our humanitarian sentiments, on the ethical injunction to care, and to manage socially to the best of our techno-managerial abilities the perversities of late capitalist post-political spatiality, or can a different politics and geographical inscription of the process of being-in-common be thought and materialized?

If both the critical socio-spatial analysis of the state of the situation, of the police order, and the call for ethical reinscription of politics fail in their emancipatory desire, thought has to be redirected toward a reinscription of the political that revolves around recentering/redesigning space as an egal-ibertarian political field of disagreement and dissensus, literally opening up space that permits acts that claim and take a place in the order of things. The key lesson to be learned from this intervention is that politics does not arise from the choreography of the social, but stands as the meeting point of the police and the political. The political configures its own theater, one that opens up a new spatiality, albeit within the given distribution of times and spaces, within the specific historical-geographical configurations and their unevenness that mark the existing socio-spatial order. It appears from within the police order, but acts at a distance from the state of the situation. The emergence of the political under the name of equality and freedom denotes a universalizing aspiration and, therefore, always operates at a certain minimal distance from the state/the police:[5] it emerges not from within the dispositive of the police, but where it is not supposed to be, and that is, in public space. The political is theatrically staged and enacted in the act of transgressing the socio-spatial configuration such that simultaneously equality is performed and the wrong (the inegalitarian practice inscribed in the police) exposed. However, the unfolding of a political procedure, in turn, articulates with and is enmeshed in the complex spatial configurations and uneven geographies that mark its eventual location. Politics occurs when the universalizing aspirations of the political event meet with the "local" specificities of the police order. While in the examples mobilized in this book certain universalizing procedures can be discerned, they always enmesh with the specific, complex, and uneven geographies that the very emergence of the political seeks to transform.

All this centers on rethinking equality politically, in other words, think-ing equality not as a sociologically verifiable concept or procedure that permits opening a policy arena that will remedy the observed inequalities (utopian/normative/moral) some time in a utopian future (i.e., the standard recipe of left-liberal urban policy prescriptions), but as the axiomatically given and presupposed, albeit contingent, condition of democracy. This requires extraordinary designs (both theoretically and materially), ones that cut through the master signifiers of consensual governance. Elements of such transgressive metonymic redesigns include thinking the spatialities of opposition/dissensus, of polemic, of agonistic (ac)count and performa-tive staging of equality.

The final and perhaps most important task is to traverse the fantasy of the elites, a fantasy that is sustained and nurtured by the imaginary of an autopoietic world, the hidden hand of market exchange that self-regu-lates and self-organizes, serving simultaneously the interests of the Ones and the All, the private and the common. The socialism for the elites that sutures the space of contemporary politics is really one that mobilizes the commons in the interests of the elite through the mobilizing and disci-plinary registers of post-democratic politics. It is a fantasy that is further sustained by a double phantasmagorical promise: on the one hand the promise of eventual enjoyment—"believe us and our designs will guar-antee your enjoyment"—that is forever postponed, that becomes a true utopia. On the other hand, there is the recurrent invocation of the specter of catastrophe and disintegration if the elites' fantasy is not realized, one that is predicated upon the relentless cultivation of fear (ecological disin-tegration, excessive migration, terrorism, economic-financial crisis, etc.), a fear that can only be managed through technocratic-expert knowledge and elite governance arrangements (see chapter 6). This fear of catastrophe has debilitating and disempowering effects—it sustains the impotence for naming and designing truly alternative places, truly different emancipa-tory spatialities and urbanities.

Traversing elite fantasies requires the intellectual and political courage to imagine the collective production of space, the inauguration of new political trajectories of living life in common, and, most important, the courage to choose, to take sides, to declare fidelity to the egalibertarian practices already prefigured in some of the evental place-moments that mark contemporary insurgencies. Most important, traversing the fantasy of the elites means recognizing that the social and ecological catastrophe that is announced every day as tomorrow's threat is not a promise, not something to come but already the Real of the present. In that sense, we have to reclaim spatial egalibertarian utopias as an utmost necessity for today.

II Practices of Post-Politicization

In part I, I explored the contested process of post-politicization and the gradual installation of post-democratic institutional regimes. In the process, a diagnostic of the deepening process of depoliticization through the establishment of techno-managerial forms of governance was provided. This, in turn, opened up a theoretical space for considering "the political difference" and interrogating the twin process of depoliticization and repoliticization. Of course, this analysis remained at the level of theoretical thought. In part II of the book, the concern shifts to considering two key domains through which post-political forms of governing have been actively constructed over the past two decades or so. Here we are less concerned with a theoretical and analytical excavation of the political condition we are in, but more with the grounded processes through which such post-democratic configuration became enacted and institutionalized. I shall consider the dialectic between depoliticization and repoliticization through the lens of the question of nature and the environment on the one hand and of the city and urbanization on the other.

4 Post-Politicizing the Environment: "Ecology as the New Opium of the Masses"[1]

Let's start by stating that after "the rights of man," the rise of the "the rights of Nature" is a contemporary form of the opium of the people. It is an only slightly camouflaged religion: the millenarian terror, concern for everything save the properly political destiny of peoples, new instruments for control of everyday life, the obsession with hygiene, the fear of death and catastrophes. ... It is a gigantic operation in the depoliticization of subjects. (Badiou 2008b, 139)

Every time we seek to mix scientific facts with aesthetic, political, economic, and moral values, we find ourselves in a quandary. ... If we mix facts and values, we go from bad to worse, for we are depriving ourselves of both autonomous knowledge and independent morality. (Latour 2004, 4)

In 2005, the *Guardian's* global edition reported how a University of Maryland scientist had succeeded in producing "cultured meat." Soon, the reporter said, "it will be possible to substitute reared beef or chicken with artificially grown meat tissue. It will not be any longer necessary to kill an animal in order to get access to its meat. We can just rear it in industrialized labs." A magical solution, so it seems, that might tempt vegetarians to return to the flock of animal protein devotees, while promising yet again (after the failed earlier promises made by the pundits of pesticides, the green revolution, and now genetic engineering and GMO products) the final solution for world hunger and a more sustainable life for the millions of people who go hungry now. Meanwhile, NASA is spending circa US$40 million annually on figuring out how to recycle wastewater and return it to potable conditions, something that would of course be necessary to permit space missions of long duration, but which would be of significant importance on Earth as well. At the same time, sophisticated new technologies are developed for sustainable water harvesting, a more rational use of water, or a better recycling of residual waters, efforts defended on the basis of the need to improve the life of 2.5 billion people who do not have adequate access to safe water and sanitation. The Breakthrough Institute keeps chipping away at advocating large-scale geo-reengineering of the earth system that would save the planet from overheating through the accumulation of greenhouse gases in the atmosphere while making sure that capitalist economic growth can not only be maintained but also accelerated as well with the help of new large-scale, ecologically sensible megatechnical geo-constructionist interventions (Ernstson and Swyngedouw 2018).

In the meantime, other "natures" keep wreaking havoc around the world. Tsunamis and hurricanes devastate homes and ecologies, forest fires blaze through Spain or California or Canada, HIV continues its genocidal march through Sub-Saharan Africa. Europe watches anxiously the nomadic wanderings of the avian flu or Zika virus and waits, almost stoically, for the moment such deadly viral diseases will travel more easily from body to body. In the UK, male life expectancy between the "best" and "worst" residential areas is now more than eleven years and the gap is widening with life expectancy actually falling (for the first time since World War II) in some areas.[2] Tuberculosis is endemic again in East London, obesity is rapidly becoming the most seriously lethal socio-ecological condition in our fat cities. And, as the ultimate cynical gesture, nuclear energy is again celebrated and iconized as the world's savior by many elites, including Gaia-warrior James Lovelock. They stage the nuclear option as the sane response to global warming while sustaining our insatiable taste for energy.

These examples and vignettes all testify to the blurring of boundaries between the human and the artificial, the technological and the natural, the nonhuman, the post-human, and the cyborg-human; they certainly also suggest that there is all manner of "natures" out there. While some of the preceding examples promise "sustainable," adaptive, or resilient forms of development, others seem to stray further away from what might be labeled as sustainable or resilient. At first glance, Frankenstein meat, cyborg waters, nuclear fission, and stem cell research are staged as possible ways to deal with serious socio-environmental problems while solving significant social problems (animal ethics and food supply on the one hand, dwindling freshwater resources or energy poverty on the other). Sustainable and adaptive or resilient solutions to our global predicament are in demand worldwide and techno-managerial fixes for our precarious environmental conditions are in feverish development. Sustainability and resilience, so it seems, is in the making.

"Responsible" scientists, environmentalists of a variety of ideological stripes and colors, together with a growing cohort of business leaders and politicians, keep on spreading doomsday and dystopian messages about the clear and present danger of pending environmental catastrophes that will be unleashed if we refrain from immediate and determined action. Particularly the threat of global warming is framed in apocalyptic terms if the atmospheric accumulation of CO_2 (which is of course the classic "side effect" of the accumulation of economic capital in the troposphere) continues unheeded (see also chapter 5). The world as we know it will come to a premature end (or be seriously mangled) unless we urgently reverse, stop, or at least slow down global warming and return the climate to its status quo ante. Political and regulatory technologies (such as those negotiated in successive Conference of the Parties (COP) to the 1992 United Nations Framework Convention on Climate Change meetings and CO_2 reducing techno-machinery (like hybrid cars, sequestration technologies, and geo-engineering projects) are developed that would, so the hope goes, stop the threatening evolution and return the earth's temperature to its apparently benevolent earlier condition. From this perspective, a resilient future is predicated upon a return, if we can, to a perceived global climatologic equilibrium situation that would permit a sustainable continuation of the present world's way of life.

So, while one sort of environmental practice seems to be predicated upon feverishly developing new natures (like artificial meat, cloned stem cells, or manufactured clean water), forcing nature to act in a way we deem sustainable or socially necessary, the other type is predicated upon limiting or redressing our intervention in nature, returning it to a presumably more benign condition, so that human and nonhuman coexistence in the medium term and long term can be assured. Despite the apparent contradictions of these two ways of "becoming sustainable" (one predicated upon preserving nature's status quo, the other predicated upon producing new natures), they share the same basic vision that techno-natural and socio-metabolic interventions, combined with appropriate good governance and management, are urgently needed if we wish to secure the survival of the planet and much of what it contains.

But these examples also show that nature is not always what it seems to be. Manufactured meat, dirty water, bird-flu virus symbiosis, stem cells, fat bodies, heat waves, tsunamis, hurricanes, genetic diversity, CO_2, to name a sampling, are radically different things, expressing radically different natures, pushing in different directions, with heterogeneous consequences and outcomes, and with radically different human/nonhuman connectivities. If anything, before we can even begin to unpack the socio-ecological conundrum we are in, these examples certainly suggest that we urgently need to interpolate our understandings of "nature," revisit what we mean by nature, and, what we assume nature to be.

Nature Does Not Exist!

More than two decades ago, Raymond Williams argued: "Nature is perhaps the most complex word in the language," wrought with all manner of histories, geographies, meanings, fantasies, dreams, and wish images (Williams 1988, 221). Yet, he also concurred that Nature is socially and politically one of the most powerful and performative metaphors of language (Williams 1980). In the wake of the current environmental crisis, Nature has gained considerable traction in political debate, economic argument, environmental conflict, and public intervention. Nature seems to be thoroughly politicized, a process further nurtured by the inauguration of the notion of the Anthropocene (Swyngedouw and Ernstson 2018). The latter indeed signals a further recognition of how both human and nonhuman natures have become an integral part of the deep-time dynamics of the earth system. If there is a political challenge in need of exploration, Nature must undoubtedly rank very highly. And this is all the more urgent as the socio-ecological conditions—the "states of nature" as it were—in many places on earth as well as globally are under serious stress. In this chapter, I shall denote "Nature" as the discursively constructed imaginaries and fantasies used to describe and capture nonhuman conditions, things, and processes, while I shall reserve "nature" or "natures" to denote specific nonhuman matter and concrete physical and (socio-)ecological processes.

Nature is indeed very difficult to pin down. Is it the physical world around and inside us, like trees, rivers, mountain ranges, HIV and Ebola viruses, microbes, elephants, oil, cocoa, diamonds, clouds, neutrons, the heart, shit, and so on? Does it encompass things like manicured roses in a botanical garden, freshly squeezed orange juice, Adventure Island in Disneyland (one of the most biodiverse ecotopes on earth), a Richard Rogers skyscraper, sewage flows, genetically modified tomatoes, and a hamburger? Should we expand it to include greed, avarice, love, compassion, hunger, death, … ? Or should we think about it in terms of dynamics, relations and relational processes like climate change, hurricane movements, species formation and species extinction, soil erosion, water shortages, food chains, plate tectonics, nuclear energy production, wormholes, supernovas, and the like?

In his book, provocatively titled *Ecology without Nature*, Timothy Morton calls Nature "a transcendental term in a material mask [that] stands at the end of a potentially infinite series of other terms that collapse into it" (Morton 2007, 14). He distinguishes between at least three places or meanings of nature in our symbolic universe. First, as a floating signifier, the "content" of Nature is expressed through a range of diverse terms that condense in the Name of Nature: olive tree, parrot fish, SARS virus, love, reproduction, the Alps, mineral water, markets, desire, profits, CO_2, money, competition, and more. Such metonymic lists, although offering a certain unstable meaning, are inherently slippery, and show a stubborn refusal to fixate meaning durably or provide consistency. Nature as metaphor remains empty (or floating); its meaning can only be gleaned from metonymic references to other, more ordinary, signifiers.

Second, Nature has "the force of law, a norm against which deviation is measured" (Morton 2007, 14). This is the sort of invocation of Nature that is mobilized, for example, to normalize heterosexuality and to think that queerness is deviant and unnatural, or that sees competition between humans as natural and altruism as a produce of culture (or vice versa). Normative power inscribed in Nature is invoked as an organizing principle that is transcendental and universal, allegedly residing outside the remit allocated to humans and nonhumans alike, but that exercises an inescapable performative effect and leaves a nonalienable imprint. This is a view that sees Nature as something given, as a solid foundational (or ontological) basis for our acting in the world and that can be invoked to provide an anchor for ethical or normative judgments of ecological, social, cultural, political, or economic procedures and practices. Consider, for example, how many of the recent efforts to nudge individual and collective activities in more sustainable directions legitimize their activities by invoking some transcendental view of a Nature that is radically out of sync and urgently requires rebalancing. Here, the functioning of Nature is offered as guiding principles on which both to found and develop a politics of the environment, one that is in tune with the rhymes and rhythms of Nature while promising a cohesive and integrated society.

And, third, Nature contains a plurality of fantasies and desires such as, for example, the dream of a sustainable nature, the desire for love making on a warm beach under the setting sun, the fear for the revenge of Nature if we keep pumping CO_2 into the atmosphere. Nature is invoked here as the stand-in for other, often repressed or invisible longings and passions—the Lacanian *objet petit a*[5] around which we shape our desires and that covers up for the lack of ground on which to base our subjectivity (Žižek 1999). This is a procedure by which we invest in Nature, displaced onto the plain of the Other, our libidinal desires and fears, a displacement of the abyss that separates the disavowed Real hard kernel of being from the symbolic world in which we dwell. It is the sort of fantasy displayed in calls for the restoration of a true (original but presumably now lost) humane social and ecological harmony by restoring the world's ecological balance. Here, Nature is invoked as the external terrain that offers the promise, if attended to properly, of finding or producing a truly happy and harmonious life (see Stavrakakis 1997).

In sum, the very uses of Nature imply simultaneously an attempt to fixate its unstable meaning while being presented as a fetishized Other that reflects or, at least, appears as a symptom through which our displaced deepest fears and longings are expressed. As such, the concept of Nature becomes ideology par excellence and functions ideologically, and by that I mean that it forecloses thought, disavows the inherent slippery of the concept, and ignores its multiplicities, inconsistencies, and incoherencies (Morton 2007, 24). Every attempt to suture, to fill in exhaustively, and to colonize the meaning of Nature is part of hegemonizing drives that are inherently political, but are not recognized as such (Laclau and Mouffe 1985; Stavrakakis 2000). Suturing Nature's meaning is of course systematically undertaken in almost all public debates, policy documents, and discourses that invoke Nature or the environment. The disavowal of the empty core of Nature by colonizing its meaning, by filling out the void, staining it with inserted meanings that are subsequently generalized and homogenized, is the gesture par excellence of depoliticization, of placing

Nature outside the political, that is outside the field of public contestation, dissensus, and disagreement while at the same time posting Nature and its functioning as a normative template for social action and policy intervention.

It is in this sense that Morton proposes "to think ecology without nature," to abandon the concept of Nature all together. This is not a silly gesture to disavow the Real of all the things, feelings, and processes associated with nonhuman stuff that I have listed. On the contrary, it is exactly the recognition of the inherent slipperiness and multiplicities of meaning suggested by such metonymic lists of really existing things, emotions, and processes that urges us to consider that perhaps the very concept of Nature itself should be abandoned.

Slavoj Žižek makes a similar point when he states controversially that "Nature does not exist!" (Žižek 1992). His Lacanian perspective insists on the difference "between series of ordinary signifiers and the central element which has to remain empty in order to serve as the underlying organizing principle of the series" (Žižek 2000, 52). Nature constitutes exactly such a central (empty) element whose meaning only becomes clear by relating it to other more directly recognizable signifiers (like cat, plant, rocks, bacteria). While every signifier is to a certain extent floating (that is undecided with respect to its associated referent), it is much easier to imagine, say, what "cat" stands for (despite the great number of different sorts of cats, let alone the infinite emotive and other meanings individuals associate with these creatures) than what Nature stands for. For Žižek too, any attempt to suture the meaning of empty signifiers is a decidedly political gesture. Yet, for him, the disavowal or the refusal to recognize the political character of such gestures, and the attempts to universalize the socially and historically situated and politically positioned meanings inscribed metonymically in Nature lead to perverse forms of depoliticization, to rendering Nature politically mute and socially neutral. In *Looking Awry*, Žižek agrees with the consensual view that the current ecological crisis is indeed a radical condition that not only constitutes a real and present danger, but one that also, equally importantly, "questions our most unquestionable presuppositions, the very horizon of our meaning, our everyday understanding of 'nature' as a regular, rhythmic process" (Žižek 1992, 34). It raises serious questions about what were long considered self-evident certainties. He argues that this fundamental threat to our deepest convictions of what we always thought we knew for certain about Nature is co-constitutive of our general unwillingness to take the ecological crisis completely seriously in a political sense. It is this destabilizing effect that explains "the fact that the typical, predominant reaction to it still consists in a variation of the famous disavowal, "I know very well (that things are deadly serious, that what is at stake is our very survival), but just the same I don't really believe, ... and that is why I continue to act as if ecology is of no lasting consequence for my everyday life" (Žižek 1992, 35). The same unwillingness to question our very assumptions about what Nature is (and even more so what natures might "become") also leads to the typical obsessive reactions of those who *do* take the ecological crisis seriously. Žižek considers the case of the environmental activist, who in his or her relentless and hysterical activism to achieve a transformation of society in more ecologically sustainable ways expresses a fear that to stop acting would lead to catastrophic consequences. In his words, hysterical acting out becomes a tactic to stave off the ultimate catastrophe, that is, "if I stop doing what I am doing, the world will come to an end in an ecological Armageddon." Others, of course, see all manner of transcendental signs in the "revenge of Nature," read it as a message that signals our destructive intervention in nature and urge us to change our relationship with Nature. In other words, we have to listen to Nature's call, as signaled by the pending environmental catastrophe, and respond to its message that pleads for a more benign, ethically inflected, associational relation with nature, a post-human horizontal and modest affective connectivity, as a cosmopolitical partner in dialogue. While the first attitude radically ignores the reality of possible ecological disaster, the other two, which are usually associated with actors defending sustainable or resilient solutions for our current predicament, are equally problematic in that they both ignore or are blind to the inseparable gap between our symbolic representations (our understandings) of Nature and the actual acting of a wide range of radically different and often-contingent natures. In other words, there is—of necessity—an unbridgeable gap, a void, between our dominant view of Nature as a predictable and determined set of processes that tends toward a (dynamic) equilibrium—but one that is disturbed by our human actions and can be "rectified" with proper sustainable practices on the one hand, and on the other hand, the acting of natures as an (often) unpredictable, differentiated, incoherent, open-ended, complex, chaotic (although by no means unordered or un-patterned) set of processes—a view that insists on uncertainty, immanence and open-endedness. The latter implies the existence not only of many natures, but, more important, it also assumes the possibility of all sorts of possible future natures, all manner of imaginable different human/nonhuman assemblages and articulations, and all kinds of different possible socio-environmental becomings.

Bruno Latour, albeit from a completely different epistemological perspective, equally proposes to ditch the concept of Nature. For Latour, there is no such thing as Nature in itself and for itself or something like Society (or Culture) (Latour 1993). For him, the collection (or imbroglios) of human and nonhuman, organic and inorganic things that compose the world consists of continuously multiplying nature-culture hybrids, or quasi-objects in Latour's terminology. With Michel Serres and others, Latour argues that these socio-natural "messy" things are made up of proliferating sets of networked socio-natural assemblages and constellations that are defined as quasi-objects; they stand between the poles of nature on the one hand and culture on the other. They are simultaneously both and neither, yet they are socio-ecologically significant and politically performative (Latour 2005). They form the socio-natures that define, choreograph, and sustain everyday lives and things. For example, think of greenhouse gases, Dolly the cloned sheep, a dam, a bottle of milk, water networks, or mobile communication waves (Swyngedouw 1996a). They are simultaneously social/cultural and natural/physical, and their coherence—that is, their relative spatial and temporal sustainability and endurance—is predicated upon assembled networks of human and nonhuman relations that form concrete socio-physical constellations (Swyngedouw 2006). This perspective, too, rejects retaining the concept of Nature and suggests in its stead to consider the infinite heterogeneity of the procedures of assembling, dissembling, and reassembling the rhizomatic networks through which things, bodies, natures, and practices become enrolled and through which relatively stable quasi-objects come into purview (Castree 2003; Braun 2006).

The world is radically heterogeneous, and the more-than-human (human and nonhuman assemblages like a cow, a personal computer, the parliament, irrigation systems, a transport network) collectives that constitute the almost infinite collection of things we call "world," "Earth," or "Cosmos" conceal relationally constituted imbroglios that "are of a highly variable duration and spatial extent—sometimes very durable, sometimes of seemingly well-bounded extent" (Henderson 2009, 284). This Latourian gesture also attempts to repoliticize Nature, to let quasi-objects enter the public assembly of contentious political negotiations and considerations. For Latour too there is nothing left to retain from the concept of Nature (Latour 2004). This hyper-constructionist view, a monist understanding of the nature-society imbroglio, also points to the end of Nature, to insist that all nature has now become fully socialized and hence subject to all manner of possibilities for reconstructing the earth as we know it (Neyrat 2016).

Despite the rejection of the concept of Nature advanced by these theorizations, it is indisputably the case that many of the world's environments are in serious ecological trouble and environmentalists, policy-makers, intellectuals, and activists desperately search or call for urgent and immediate action in the face of the clear and present danger posed by environmental degradation and possible ecological collapse. For those attuned to the mess many of the world's environments are in, linguistic acrobatics as those introduced in this chapter might sound esoteric at best and nonsensical and counter-productive at worst, coming from the usual suspects of critical social theorists and attractive, but politically vacuous and pragmatically impotent cultural musings.

Interestingly, however, some "hard" scientists seem to echo these critical social theory perspectives, albeit with a slightly different terminology and from an altogether different vantage point. The exemplary work of Levins and Lewontin (Levins and Lewontin 1985; Lewontin and Levins 2007), Harvard University biologists and ecologists, comes to strikingly similar conclusions, yet does so from a Marxist dialectical materialist perspective. They too agree that Nature has been filled in by scientists with a particular set of universalizing meanings that ultimately depoliticize Nature, and evoke a series of distinctly ideological principles that facilitated particular mobilizations of such "scientifically" constructed Nature. While eighteenth- and nineteenth-century scientific views of nature were enthralled by notions of change, revolution, and transformation, twentieth-century biology, Levins and Lewontin argue, settled Nature down, and reduced it to a homeostatic constellation. As they put it: "We are at the End of Natural History. The world has settled down, after a rocky start, to a steady state. Constancy, harmony, simple laws of life that predict universal features of living organisms, and self-reproduction and absolute dominance of a single species of molecule, DNA, are the hegemonic themes of modern biology" (Lewontin and Levins 2007, 13–14).

Of course, modern biology does not reject the radical environmental transformations affecting our living environments. However, unforeseen changes are seen either as the effect of "externalities," in other words, humans' irresponsible intervention in the steady state/evolution of a mechanical nature, or as catastrophic turbulence resulting from initial relations that spiral out in infinitely complex and greatly varying configurations such as those theorized by chaos theory or complexity theory. While the former insists on nature's innate stabilizing force disrupted by external (human) agency, the latter reduces the complex vagaries of environmental change to the unpredictable outcome of forces immanent in apparently infinitesimal original conditions. Both perspectives deny that the biological world is inherently relationally constituted through contingent, historically produced, infinitely variable forms in which each part, human or nonhuman, organic or nonorganic, is intrinsically bound up with the wider relations that make up the whole.[3] Levins and Lewontin abhor a simplistic, reductionist, teleological, and ultimately homogenizing view of Nature. They too insist that a singular Nature does not exist, that there is no transhistorical and/or transgeographical transcendental natural or original and foundational state of things, of conditions or of relations, but rather there is a range of different historical natures, relations, and environments that are subject to continuous, occasionally dramatic or catastrophic, and rarely, if ever, fully predictable changes and transformations. Their dynamics are shaped by the time–space-specific relational configurations in which each part is inserted. Neither these parts nor the totality of which they are part can be reduced to a foundational given (whether "mechanical" or "chaotic"). They eschew such expressions as "it is in the nature of things" to explain one or another ecological or human behavior or condition. They hold to a relational and historically contingent view of biological differentiation and evolution. The world is in a process of continuous becoming through the contingent and heterogeneous recompositions of the almost infinite (socio-)ecological relations through which new natures come into being. They see the relations of parts to the whole and the mutual interaction of parts in the whole as the process through which both individuals and their environments are changed (see also Harvey 1996). In other words, both individuals and their environments are coproduced and coevolve in historically contingent, highly diversified, locally specific and often not fully accountable manners. For Levins and Lewontin, therefore, no universalizing or foundational claim can be made about what Nature is, what it should be or where it should go. Nature does not exist for them either.

This is also the view shared by the late evolutionary biologist Stephen Jay Gould who saw evolution not as a gradual process, but one that is truncated, punctuated, occasionally catastrophic and revolutionary but, above all, utterly contingent (Gould 1980). There is no safety in Nature—Nature is unpredictable, erratic, moving spasmodically and blind. There is no final guarantee in Nature on which to base our politics or the social, on which to mirror our dreams, hopes, or aspirations, on which to ground our dreams and plans for a different, let alone better, and socio-ecologically more sensitive mode of living together. To put it bluntly, bringing down (or not, as the case may be) CO_2 emission does affect the global climate and shapes socio-ecological patterns in distinct manners (that are of course worthy of both scientific exploration and ethical concern); such process, even if successful, would not produce in itself the "good" society in a "good" environment.

Both the critical cultural perspectives and post-foundational evolutionary views explored here lead to a series of arguments and claims about Nature and how to think, conceptualize, and/or politicize it. This articulation between a nonfoundational view of nature and of politics/the political is the conceptual conundrum I wish to disentangle further in this chapter. Dissecting the conceptual challenges posed by the mobilization of Nature in a wide range of social sciences, political discourses, and policy or managerial practices is absolutely vital in a world in which post-democratization dovetails with socio-ecological dynamics such as resource depletion, climate change, or environmental degradation posing challenges that, if unheeded, might possibly lead to the premature end of civilization as we know it, to the end of us before our sell-by date has expired.

The main points of argument I wish to unfold in this chapter are as follows:

1. Nature and its more recent derivatives, like "environment" or "sustainability" are "empty" signifiers.
2. There is no such thing as a singular Nature around which an environmental policy, a more sustainable socioeconomic future, or an environmentally sensitive policy and planning can be constructed and performed. Rather, there is a multitude of natures and a multitude of existing, possible, or practical socio-natural relations.

3. The obsession with a singular Nature that requires sustaining or, at least, managing is sustained by a particular quilting of Nature that forecloses asking political questions, deepens the depoliticizing dynamics associated with post-politicization, and renders mute the political debate about immediately and very possible alternative socio-natural arrangements.

4. I conclude with a call for a politicization of the environment, one that is predicated upon the recognition of the indeterminacy of nature, the constitutive split of the people, the unconditional democratic demand of political equality, and the real possibility for the inauguration of different possible public socio-ecological futures that express the democratic presumptions of freedom and equality.

The Empty Core of Nature: Multiple Natures

As suggested in this chapter, it is difficult, if not impossible, to define exactly what Nature is. Every attempt to nail down or to fix its meaning seems futile at best and politically problematic at worst; Nature's content is like an eel, slipping away the very moment you think you finally grasped it. Nature is an empty or floating signifier. Empty signifiers gain a certain, yet unstable, contingent, contestable, and invariably temporary coherence or content (but in the process, they are simultaneously emptied out of a determinate meaning—they are rendered floating) through the mobilization of a metonymic list, a chain of equivalences or equivalent signifiers that quilt their meaning (Žižek 1989; Stavrakakis 1997). The longer the list of signifiers (like fish, rain, orgasm, earthquake, evolution, skin pigment, greed) that have to be strung together to give a concept like Nature some sort of meaning or content, the more contested, indeterminate, and inchoate the concept becomes. Nature becomes a tapestry, a *montage*, of meaning, held together with quilting points (like the upholstery of a Chesterfield sofa). For example, in today's environmental parlance, biodiversity, eco-cities, pollutants, and CO_2 can be thought of as quilting points through which certain meanings of Nature are knitted together. Moreover, these quilting points are also more than mere anchoring points; they refer to a beyond of meaning, a certain enjoyment that becomes structured in fantasy (in this case, an environmentally sound and socially harmonious order).[4]

This emptying out of a fixed meaning of Nature has been a systematic feature of late modernity, particularly as signifying chains of what Nature "really" is multiplied in parallel to the proliferation of both scientific controversies and socio-political, cultural, or other differentiations. We shall briefly explore this multiplication of narratives of Nature. Consider, for example, how in premodern times Nature was—in the West at least—signified through a divine order, God's creation, whereby Nature was relegated to the domain of the transcendental. Nature and God were interchangeable, offering a meaning of Nature that gained content through its relationship with a world order understood as divine and beyond the mortal. With the advent of enlightenment and early modernity, Nature became quilted through referents such as science, rationality, truth, and clockwork mechanics. A new Truth of Nature became gradually established, one of a singular Nature that behaved mechanically, and through the mobilization of the correct application of the technologies of rationality and proper scientific method, its law-like operation could be deciphered, coded, and recoded, and subsequently manipulated to serve human ends. Nature possessed an internal logic, and teleology, a mode of arranging itself that was self-contained and self-organizing, one that did not require a referent like God or Man; Nature's inner logic was self-generating and self-referential, one that defied an articulation with either the divine or the human. It was a view of Nature that rapidly asserted itself as the true view against all lingering superstitious remainders, whether pagan or Christian. It was also a view that increasingly considered Nature as distinct and separate from the constructed social and cultural world of human interactions (Smith 2008b). This ultimately pitted Nature against civilization, the natural against, or at least as separate from, the social, the political, and the cultural.

Yet, this scientific notion of Nature began to explode rapidly from the nineteenth century onward. While the Nature/Science/Rationality imbroglio remained firmly in place and consolidated itself, primarily through the increasingly successful development of the natural sciences and its mesmerizing applications in all domains of life, from the nuclear bomb to the dishwashing machine, a plethora of other signifying chains began to quilt Nature's meaning. Take, for example, the Romanticism of nineteenth-century notions of Nature. At a time during which the frontiers of nature (in the sense of the external dehumanized nature that Enlightenment thinking had framed) receded through the spiraling mobilization of an expanding array of nonhuman things in capitalist production and consumption on the one hand and colonial-imperial exploration and incorporation of "new" lands and natures in a rescaled eco-political orbit on the other, Nature (including the "noncivilized" peoples it contained like Native Americans and enslaved/colonized Africans) became associated with untamed wilderness, (lost) originality, moral superiority (against the moral decay of the "civilized" world), utopian idyllic Arcadia, and sublime awesome beauty. Or consider the signifying chain that emerged with the first signs of ecological crisis in late nineteenth-century cities, something that was particularly prevalent in rapidly urbanizing societies where the deteriorating sanitary conditions, the bacteriological infestations, the metabolic rift produced by the separation of town and countryside, and so on, opened up a tremendous new real and symbolic space for Nature. From then onward, Nature would also be partly symbolized as dangerous and threatening, as fearful—yet man-made—in its urban manifestations (Kaika 2005; Gandy 2006).

In sum, modernization produced a cacophony of metonymic lists associated with Nature, none of which exhausted the vagaries, idiosyncrasies, and heterogeneous acting of the different and changing forms of the nonhuman that composes the world. The inclusion of post-colonial, post-human, and other forms of knowledge in recent years has expanded the range of possible meanings inscribed in Nature. Moreover, these forms proliferated as the number of socio-natural things, these hybrids of humans and of nature—what Bruno Latour calls "quasi-objects" (Latour 1993) or Donna Haraway calls "cyborgs" (Haraway 1991)—multiplied with the intensifying assembling of human and nonhuman processes (as in, for example, nuclear energy, the making of trans-uranium elements, genetic manipulation, techno-natural constructs like water systems, high-voltage energy lines, megacities, and the like).

In recent years, and in particular as a result of the growing global awareness of "the environmental crisis," the inadequacy of our symbolic representations of Nature became more acute again as the Real of natures, in the form of a wide variety of ecological threats (global warming, new diseases, biodiversity loss, resource depletion, pollution) invaded and unsettled our received understandings of Nature, forcing yet again a transformation of the signifying chains that attempt to provide content for Nature, while at the same time exposing the impossibility of capturing fully the Real of natures (Žižek 2008b). Ecologists (deep or otherwise), environmental modernists, post-materialists, a diversity of environmental movements, earth systems scientists, corporations concerned with the dwindling of resources on which their growth and profits are based, new insights generated by a still-successful and sprawling natural science (but one somewhat more sensitive to ethical issues after the backlash unleashed by the perverse "successes" of the nuclear age), even the captains of industry and a new generation of political elites began to extend, transform, or reinvent the arsenal of meanings assigned to Nature.

These new imaginings of what Nature is revolve around the emergence and consolidation of complexity theory and theories of nonlinear dynamics that cherish "emergence," "resilience," the "indeterminacy" of nature, and radical openness, but do so without explicit attention to its parallel deployment in political economy, culture, and politics (for a discussion, see Hornborg 2009 and Nadasdy 2007). As Bruce Braun insisted, in his careful dissection of the historiographies of the new materialisms, it is not difficult to discern the parallels between nondeterministic geo-sciences (including complexity science and "resilience" studies) and the varieties of new materialisms associated with more-than-human and object-oriented ontologies (Braun 2015; see also Protevi 2013). Both rose to prominence—but in a contingent manner—in the context of the deep crisis of capitalism in the 1970s and its attempts to search for a fix to the malaise in the process of neo-liberalization (see Walker and Cooper 2010; Nelson 2015). What is at stake here is precisely how the promise of a fast-forwarding capitalist modernization can proceed, despite or perhaps because of, an altered ontological premise, and with a different storyline to mask what is really at stake.

The point of the preceding argument is that the natures we see and work with are necessarily imagined, scripted, and symbolically charged as Nature. These inscriptions are always inadequate; they leave a gap or a stain that escapes symbolization, a hard kernel, a remainder, and maintains a certain unbridgeable distance from the natures that are there, which are complex, chaotic, often unpredictable, radically contingent, historically and geographically variable, risky, patterned in endlessly complex ways, ordered along strange attractors (see, for example, Prigogine and Stengers 1984). In other words, there is no Nature out there that needs or requires salvation in name of either Nature itself or a generic Humanity. There is nothing foundational in Nature that needs, demands, or requires sustaining. The debate and controversies over Nature and what to do with it, in contrast, signal rather our political inability to engage in directly political argument and strategies about rearranging the socio-ecological coordinates of everyday life, the production of new socio-natural configurations, and the arrangements of socio-metabolic organization (something usually called capitalism) that we inhabit. In the next section, we shall exemplify and deepen further this conceptual and theoretical analysis by looking at the depoliticizing notion of sustainability and sustainable development, symptomatic concepts that have become the hegemonically and consensually agreed metaphors to signal the ecological quandary we are in and to chart a prophylactic course that might prevent a further acceleration of our descent into a socio-ecological abyss. Indeed, one of the key signifiers that have emerged as the pivotal empty signifier to capture the growing concern for a Nature that seemed to veer off balance is, of course, sustainability.

The Fantasy of Sustainability: A "Resilient" Planet

Although there may be no Nature, there certainly is a politics of nature or a politics of the environment. The collages of apparently contradictory and overlapping vignettes of the environmental conditions outlined earlier share one common threat that my son and Greenpeace, Oxfam and the World Bank, Barack Obama and Angela Merkel, and most of the rest of us agree on. The world is in environmental trouble. And we need to act politically now. There is indeed a widespread consensus that the earth and many of its component parts are in an ecological bind that may short-circuit human and nonhuman life in the not too distant future if urgent and immediate action to retrofit nature to a more benign equilibrium is postponed for much longer. Irrespective of the particular views of Nature held by different individuals and social groups, consensus has emerged over the seriousness of the environmental condition and the precariousness of our socio-ecological balance. BP (British Petroleum) has rebranded itself as "Beyond Petroleum" to certify its environmental credentials, Royal Dutch Shell plays to a more eco-sensitive tune, eco-warriors and philanthropists of various political or ideological stripes and colors engage in direct action in the name of saving the planet. New Age post-materialists join the chorus that laments the irreversible decline of ecological amenities, eminent scientists enter the public domain to warn of possible ecological catastrophe, politicians try to outmaneuver each other in brandishing the ecological banner, and a wide range of policy initiatives and practices, performed under the motif of sustainability or resilience are discussed, conceived, and implemented at all geographical scales. Indeed, ecological matters are elevated to the dignity of a global humanitarian cause. While there is certainly no agreement on what exactly Nature is and how to relate to it, there is a virtually unchallenged consensus over the need to be more environmentally sustainable if disaster is to be avoided.

In this consensual setting, environmental problems are generally staged as universally threatening the survival of humankind, announcing the premature termination of civilization as we know it. The discursive matrix through which the contemporary meaning of the environmental condition is woven is one quilted systematically by the continuous invocation of fear and danger, the specter of dystopian ecological annihilation, or at least seriously distressed socio-ecological conditions in the near future. Fear is indeed the crucial node through which much of the current environmental narrative is threaded, and that continues to feed the concern with sustainability (Swyngedouw 2013a). This scripting of Nature permits and sustains a post-politicizing arrangement sutured by fear and driven by a concern to manage things so that we can hold on to what we have. It is precisely this view that leads Alain Badiou to insist that ecology has become the new opium of the masses, replacing religion as the axis around which our fear of social disintegration becomes articulated. Such ecologies of fear ultimately conceal, yet nurture, a conservative or at least reactionary message. While clouded in rhetoric of the need for radical change in order to stave off immanent catastrophe, a range of technical, social, managerial, physical, and other measures have to be taken to make sure that things remain the same, that nothing really changes, that life (or at least our lives) can go on as before. This sentiment is also shared by Frederic Jameson when he claims that "it is easier to imagine the end of the world than it is to imagine the end of capitalism" (Jameson 2003, 76).

In the call for a rebalanced environmental condition, often pursued through futuristic Promethean geo-engineering proposals like large-scale solar radiation deflection panels, deep geological depositing of CO_2, or ocean fertilization technologies (see The Royal Society 2009; Shellenberger and Nordhaus 2007), many actors with very different and often antagonistic cultural, economic, political, or social positions, interests, and inspiration can find common cause in the name of a socially disembodied humanity. Al Gore's emblematic documentary, *An Inconvenient Truth*, becomes strangely enough a very convenient one for those who believe civilization as we know it (I prefer to call this "capitalism") needs to be preserved and immunized against potential calamity and revolutionary change. It calls for the rapid deployment of a battery of innovative environmental technologies, eco-friendly management principles, and resilient organizational forms (something capitalist dynamics and relations excel in producing providing they conform to the profit principle [see Buck 2007]), so that the existing socio-ecological order really does not have to change radically. Consider, for example, how the real and uncontested accumulation of CO_2 in the atmosphere, the threat of peak-oil, the expanding demands for energy and other resources, are today among the greatest concerns facing large companies. Of course, BP, ExxonMobil, IBM and other multinationals are rightly worried that the model of unbridled capital accumulation upon which their success in the twentieth century rested so crucially may face ecological constraints if not limits. This testifies to the quest of leading companies for assuring the right techno-administrative arrangements can be invented and devised to permit consumers to buy us all out of the ecological pickle we're in. Provided the correct techno-administrative apparatuses can be negotiated (like Agenda 21, carbon markets and off-setting, recycling schemes, eco-friendly hard and soft technologies, and biodiversity preservation management), the socio-ecological order as we know it can be salvaged, rescued from ecological Armageddon. It is ironic how in the context of climate change arguments, for example, technological fixes, searching for more energy-effective and fossil-fuel-replacing technologies, combined with market-conforming policy arrangements stand to guarantee that the socio-ecological order we inhabit can be sustained for some time longer, that the Holocene really can become an Anthropocene, the age of man.

The container signifier that encapsulates these post-political attempts to deal with Nature is, of course, sustainability (Gibbs and Krueger 2007). Even more so than the slippery and floating meanings of Nature, sustainability is the empty signifier par excellence. It refers to nothing and everything at the same time. Its prophylactic qualities can only be suggested by adding specifying metaphors. Hence, the proliferation of terms such as sustainable cities, sustainable planning, sustainable development, sustainable forestry, sustainable transport, sustainable regions, sustainable communities, sustainable yield, sustainable loss, sustainable harvest, sustainable resource (fill in whichever you fancy) use, sustainable housing, sustainable growth, sustainable policy. The gesture to "sustainability" already guarantees that the matter of Nature and the environment is taken seriously, that those in charge take into account our existential fears, and that "homeland security" is in good hands.

The fantasy of imagining a benign and sustainable Nature avoids asking the politically sensitive, but vital, question as to what kind of socio-ecological arrangements and assemblages we wish to produce, how this can be achieved, and what sort of environments we wish to inhabit, while at the same time acknowledging the radical contingency and undecidability of Nature. This is the clearest expression of the structure of fantasy in the Lacanian sense. While it is impossible to specify what exactly sustainability is all about (except in the most general or generic of terms), this void of meaning is captured by a multiplying series of fantasies, of stories and imaginaries that try to bridge the constitutive gap between the indeterminacies of natures on the one hand (and the associated fear of the continuous return of the Real of nature in the guise of ecological disasters like droughts, hurricanes, floods) and on the other hand, the always frustrated desire for some sort of harmonious and equitable socio-ecological living, one that disavows the absence of a foundation for the social in a Nature that, after all, does not exist. Sustainability or more precisely the quilting points around which its meaning is woven, is the environmental expert's and activist's *objet petit a*, the thing around which desire revolves, yet simultaneously stands in for the disavowed Real, the repressed core, the state of the situation (i.e., the recognition that the world is in a mess and needs drastic and dramatic, that is revolutionary, action). It is in this phantasmagoric space that the political dimension disappears to be replaced by a consensually established frame that calls for techno-managerial action in the name of humanity, social integration, the earth and its human and nonhuman inhabitants, all peoples in all places.

The Depoliticized Politics of Sustainability: An Immuno-biopolitical Fantasy

We still have to account for the formidable performativity of the notion of sustainability and its discursive success as a signifier that is popular and scientific, potentially radical yet utterly reactionary. As suggested earlier,

the theory and practices of sustainability, adaptation, and resilience provide for an apparently immunological prophylactic against the threat of an
irremediable socio-ecological disintegration. Roberto Esposito's reworking
of Foucault's analysis of bio-political governmentality, enhanced by Fréderic Neyrat's psychoanalytical interpretation, may begin to shed some light
on this deadlock (Esposito 2008, 2011; Neyrat 2010). Esposito's main claim
expands on Michel Foucault's notion of biopolitical governmentality as the
quintessential form of modern liberal state governance by demonstrating
how this biopolitical frame is sutured by an immunological drive, a mission
to seal off objects of government (the population) from possibly harmful
intruders and recalcitrant or destabilizing outsiders that threaten the bio-
happiness, if not sheer survival, of the population, and guarantees that life
can (continue to) be lived. Immunological has to be understood here as the
suspension of the obligation of mutual communal gift-giving, a form of
asylum that suspends one's obligation to participate in the rights and obligations of the commons, of the community.[6] The (neo)liberal injunction
of individual freedom and choice is precisely the founding gesture of such
an immunological bio-politics, in other words, the ring-fencing of the fragmented body from its insertion in the obligations and violence that bonds
community life (Brossat 2003).

Immuno-politics are clearly at work, for example, in hegemonic Western
practices around immigration or international terrorism. A rapidly expanding arsenal of soft and hard technologies is put in place in an ever denser
layering of immunological, technical, infrastructural, and institutional-
legal dispositives—from tighter immigration law and continuous surveil-
lance to the actual construction of steel and concrete walls and barriers,
and the proliferation of all sorts of camps and other militarized or policed
enclosures. Similar examples can be identified in the strict cordoning off
when infectious diseases threaten to spatialize in manners that could pen-
etrate the immuno-engineered eco-topian bubbles of the elite's local life. Is
it not the case that much of the sustainability and eco-managerial practices
that populate resilient socio-ecological interventions and sustainable technologies and governance practices are precisely aimed at reinforcing the
immunological prowess of the immune system of the body politic against
recalcitrant, if not threatening, outsiders (like CO_2, waste, bacteria, refugees,
viruses, ozone, financial crises, and the like) so that life as we know it can
continue (Neyrat 2008)? As Pierre-Oliver Garcia puts it: "An immunitary

power takes control of the risks, dangers and fragilities of individuals to make them live in a peaceful manner while obscuring any form of dissensus" (Garcia 2015, 321; my translation).

While the immuno-biopolitical gestures associated with refugees, (bio-) security, and economic-financial collapse customarily succeed in translocating, while nurturing, fear of potential collapse and disintegration to the terrain of a crisis to be governed or a situation to adapt to, the immuno-biopolitical dispositive is rapidly disintegrating in the face of the actually existing combined and uneven ecological catastrophe. Indeed, with respect to the ecological condition, the standard technologies of neoliberal governance, which sustain and nurture the immuno-biopolitical desire that Esposito points to as the primary logic of neoliberal governmentality, become increasingly ineffective. Is it not the case that the immuno-biopolitical managerial tactics of adaptive climate policies, sustainable socio-technical practices, and other eco-governance arrangements leave an uncanny remainder? Are we not left with the gnawing feeling that, despite the elevation of the ecological condition to the dignity of a global public concern, the socio-ecological parameters keep eroding? In spite of the combination of market-led adaptation and mitigation strategies that were argued to provide a safety wall against further climate change, for example, the Real of the ecological disintegration still gallops forward. The threat of a modified, but still presumably externally operating and increasingly monstrous nature, intensified to such an extent that the mainstream consensual view (reflected in as diverse writings as the successive reports of the Intergovernmental Panel on Climate Change (IPCC), Al Gore's crusade, or Naomi Klein's desperate call for radical change) increasingly concedes that such nature's extra-human acting may produce effects inimical to long-term human survival. While other risks (economic, refugee, or geopolitical/security crises) are subject to immuno-biopolitical gestures that promise life unencumbered in the face of potentially lethal threats by means of deepening immunological management, screening, pervasive policing, and techno-shielding, the environmental biopolitical masquerade—invariably captured by empty signifiers of sustainability, adaptation, resilience, or retro-eco-engineering—secures at best a palliative for temporary relief. The adaptive scenarios, such as for example carbon trading and carbon storage, or the desperate search for carbon-substituting soft-energy sources to deal with our environmental and climatic woes still keep nature at arm's length.

The ecological and climate conundrum remains firmly the place that hides consequences from us and from where danger emanates, no matter the range of palliatives put at our disposition. The woefully inadequate and plainly failing attempts to immunize life from the threat of ecological collapse cannot any longer be ignored.

The insistent intrusion of the Real of socio-ecological destruction undermines terminally this immunological fantasy script, exposes its unstable core, uncovers the gap between the Symbolic and the Real, and undermines its supporting discursive matrix, thereby threatening the coherence of the prevalent socio-ecological order. The incessant return of the Real of ecological disintegration might fatally undermine our drive's primordial energy as we are increasingly caught up in the horrifying vortex of radical and irreversible socio-ecological disintegration. The fantasy of eternal life meets the Real of its premature end. A radical reimagination of the socio-geo-physical constellation of the earth system is therefore urgently called for, barring the unbearable Reality of an untimely death that is now firmly on the horizon. The uncanny feeling of anxiety that all is not as it should be, that keeps gnawing, is sublimated and objectified in *objet a*, the horrifying "thing" around which both fear and desire become articulated; it is a fear that is vested in an outsider that threatens the coherence and unity of our life-world.

As Esposito argues, the immunological-biopolitical dispositive turns indeed into a thanatopolitics. In the excessive acting of the immunological drive, the dispositive turns against that which it should protect. It becomes self-destructive in a process of auto-immunization. The very mechanisms that permitted biopolitical governance in the twentieth century—the thermocene of unbridled carbon metabolization upon which capitalist expansion rested—turned into an auto-destructive process. This auto-immunization, in turn, isolates the pathological syndrome and treats it as an externalized "bad" that requires isolation and sequestration (Garcia 2015, 352–253). In other words, the mechanisms that make and secure life end up threatening its very continuation. This infernal dialectic, Fréderic Neyrat argues, is predicated upon the fantasy of absolute immunization, the fact that despite knowing very well we shall die, we act and organize things as if life will go on forever (Neyrat, Johnson, and Johnson 2014). It is precisely in this fantasy space, sustained by a human exceptionalism as the sole species capable of preventing its own death, that both the modest and

more radical imaginaries of sustainable futures find their ultimate ground
(Neyrat 2014). Indeed, the pursuit of socio-technical sustainability promises to cut the unbearable deadlock between immuno- and thanato-politics
without really having to alter the trajectory of socio-ecological change. In
fact, it deepens it. In psycho-analytical terms, the immuno-biopolitical prophylactic dispositive circulates around the death-drive, the obsessive pursuit of desire that permits covering up the inevitability of death; it is the
process that makes sure that we can go on living without staring the Real
of eventual (ex-)termination in the eye.[7] While the pursuit of happiness lies
in avoiding pain, the death drive, sustained by desire, propels us forward
as if we would live forever irrespective of (and even moved along by) the
threats we encounter on our journey to the end. The energy of the drive
is fueled by the disavowal of a certain death. It is the hysterical position
that guarantees that death remains an obscure and distant impossibility.
We can survive and do so without the necessity of facing political actions
and radically different political choices. A shift in the techno-managerial
apparatuses will suffice.

It is precisely at a time that the Real of the excessive acting of an externalized threat, in particular in the form of CO_2, can no longer be contained
and ignored that a widening and intensification of the immunological biopolitical drive is called for (Neyrat 2014). This gesture, once again, con-
fidently projects our survival into eternity without considering the need
or potential for a transformation of socio-natural relations themselves; it
invites and nurtures techno-managerial adaptions to assure the sustainability of the existing.

Conclusion

In this chapter, I have argued that the fantasy of sustainability underpins
the formation of a post-political consensual order and poses real challenges
with respect to the fundamental socio-ecological problems we face. While
Nature and sustainability do not exist outside the metonymic chains that
offer some sort of meaning, there are of course all manner of existing, possible, or desirable environments and assemblages of socio-natural relations.
Environments are specific historical results of socio-physical processes
(Heynen, Kaika, and Swyngedouw 2006). All socio-spatial processes are
indeed invariably also predicated upon the circulation, the metabolism, and

the enrolling of social, cultural, physical, chemical, or biological processes, but their outcome is contingent, often unpredictable, immensely varied, risky. These metabolisms produce a series of both enabling and disabling socio-environmental conditions. Indeed, these produced milieus often embody contradictory tendencies. Processes of socio-ecological change are, therefore, never socially or ecologically neutral. For example, the unequal ecologies associated with uneven property relations, the commodification of all manner of natures, the impoverished socio-ecological life under the overarching sign of the commodity and of money in a neoliberal order, and of the perverse exclusions choreographed by the dynamics of uneven eco-geographical development at all scales suggest how the production of socio-ecological arrangements is always a deeply conflicting, and hence irrevocably political, process. All manner of social power geometries shape the particular social and political configurations as well as the environments in which we live. Therefore, the production of socio-environmental arrangements implies fundamentally political questions, and has to be addressed in political terms. The question is to tease out who gains from and who pays for, who benefits from and who suffers (and in what ways) from particular processes of metabolic circulatory change. These flows produce inclusive and exclusive ecologies both locally and in terms of the wider uneven socio-ecological dynamics and relations that sustain these flows. Democratizing environments, then, become an issue of enhancing the democratic content of socio-environmental construction by means of identifying the strategies through which a more equitable distribution of social power and a more inclusive mode of producing natures can be achieved. This requires reclaiming democracy and democratic public spaces (as spaces for the enunciation of agonistic dispute) as a foundation for and condition of possibility for more egalitarian socio-ecological arrangements, the naming of positively embodied ega-libertarian socio-ecological futures that are immediately realizable. In other words, egalitarian ecologies are about demanding the impossible and realizing the improbable.

5 Hotting Up: Climate Change as Post-Politicizing Populism

Hammer: "If the place isn't hotting up, we're fucked!"
Beard: "Here's is the good news. The UN estimates that already a third of a million people a year is dying from climate change. Bangladesh is going down. ... Methane is pouring out of the Siberian permafrost. There is a meltdown under the Greenland ice sheet. ... Two years ago we lost forty per cent of the Arctic summer ice. ... The future has arrived, Toby."
Hammer: "Yeah, I guess."
Beard: "Toby, listen. It is a catastrophe. Relax."
(McEwan 2010, 216–217)

Climate Change and Post-Politicization

A specter is haunting the entire world: but it is not that of communism. ... Climate change—no more, no less than nature's payback for what we are doing to our precious planet—is day by day now revealing itself. Not only in a welter of devastating scientific data and analysis but in the repeated extreme weather conditions to which we are all, directly or indirectly, regular observers, and, increasingly, victims. (Levene 2005)

This chapter interrogates the relationship between two apparently disjointed themes: the consensual presentation and mainstreaming of the global problem of climate change that presents a clear and present danger to civilization as we know it unless urgent and immediate remedial immuno-biopolitical action is undertaken on the one hand, and the debate in political theory, explored in earlier chapters, that centers on the emergence and consolidation of a post-political and post-democratic condition on the other hand. While the former insists on the urgency of politically mediated actions to retrofit the global climate to a more benign and stable condition or, at least, on the need to engage in interventions that might mitigate some of its more dramatic socio-spatial and ecological consequences by means of adaptation strategies or building resilience, the latter argues that the last few decades have been characterized by deepening processes of depoliticization marked by the increasing evacuation of the political dimension from the public terrain as technocratic management and consensual policy-making have sutured the spaces of democratic politics.

I shall proceed in four steps. First, I briefly outline the basic contours of the argument and its premises. In the second section, I discuss the ways in which the present environmental conundrum is expressed via the vantage point of the climate change debate. Third, I turn to exploring how the specific staging of climate change and its associated policies is sustained by decidedly populist gestures. In the final section, I argue that this particular choreographing of climate change is one of the arenas through which a post-political frame and post-democratic political configurations have been mediated. Throughout, I maintain that current hegemonic climate change policies ultimately contribute to reinforcing depoliticization and the socio-political status quo rather than, as some suggest and hope, offering a wedge that might contribute to achieving socio-ecologically more egalitarian transformations (Castree 2009). The chapter concludes with an appeal to rethink the political, to reestablish the horizon of democratic environmental politics.

The Argument

Socialism or Barbarism.
—Karl Marx

Kyoto Protocol or the Apocalypse.
—Green saying

The argument advanced here attempts to interrogate the apparently paradoxical condition whereby the climate is politicized as never before while a group of increasingly influential political philosophers and theorists insists that the post-politicization of the public sphere (in parallel and intertwined with processes of neoliberalization) have been key markers of the political process over the past few decades. A widespread consensus has emerged over the state of nature and the precarious environmental conditions the world is in and that may lead to the premature end of civilization as we know it (Intergovernmental Panel for Climate Change 2007). The environmental condition and, in particular, global climate change are increasingly staged as signaling a great danger, of epic dimensions, that, if unheeded, might radically perturb, if not announce the premature end, of our civilization before its sell-by date has passed. The imminent danger to the future of our common human and nonhuman world calls for radical changes in all manner of domains, from the way we produce and organize the transformation and socio-physical metabolism of nature to routines and cultures of consumption (Giddens 2009). A fragile consensus on both the "nature" of the problem and the arrays of managerial and institutional technologies to mitigate the most dramatic consequences has been reached, despite differences of view, opinion, or position (see Hulme 2009). This consensus is now largely shared by most political elites from a variety of positions, business leaders, activists, and the scientific community. The few remaining skeptics are increasingly marginalized as either maverick hardliners or conservative bullies.

This elevation of climate change and its consequences onto the terrain of public concern and policy has unfolded in parallel to the consolidation of a political condition that has evacuated dispute and disagreement from the spaces of public encounter to be replaced by a consensually established frame defined as post-democratic or post-political (see chapters 2 and 3). What I am concerned with in this chapter is to consider how the environmental question in general—and the climate change argument in particular—has been and continues to be one of the markers through which post-democratization is wrought and can be mobilized as a lens through which to grapple with the contested formation of a post-politicizing frame. The politics of climate change are not just expressive of such post-democratic organization, but have been among the key arenas through which the process of post-politicization is forged, configured, and entrenched. This process of depoliticization, which operates through elevating the state of nature onto the public terrain in thoroughly depoliticized ways, calls for a reconsideration of what the political is, where it is located, and how the democratic political can be recaptured.

The Desire of the Apocalypse and the Fetishization of CO_2

If we do nothing, the consequences for every person on this earth will be severe and unprecedented—with vast numbers of environmental refugees, social instability and decimated economies: far worse than anything which we are seeing today. ... We have 100 months left to act.

—Prince Charles, March 2009; see also New Economics Foundation, http://www.neweconomics.org/content/one-hundred-months

We shall start from the attractions of the apocalyptic imaginaries that infuse the climate change debate and through which much of the public concern with the climate change argument is sustained. The distinct millennial-ism discourse around the climate coproduced a widespread consensus that the earth and many of its component parts are in an ecological bind that may short-circuit human and nonhuman life in the not too distant future if urgent and immediate action to retrofit nature to a more benign equi-librium is postponed for much longer. Irrespective of the particular views of Nature held by different individuals and social groups, consensus has emerged over the seriousness of the environmental condition and the pre-cariousness of our socio-ecological balance (Swyngedouw 2009a).

This consensual framing of the problem is itself sustained by a particu-lar scientific discourse.[1] The complex translation and articulation between what Bruno Latour would call matters of fact versus matters of concern has been thoroughly short-circuited (Latour 2004). The changing atmospheric composition, marked by increasing levels of CO_2 and other greenhouse gases in the atmosphere, is largely caused by anthropogenic activity, pri-marily (although not exclusively) as a result of the burning of fossilized or captured CO_2 (in the form of oil, gas, coal, wood) and the disappearance of CO_2 sinks and their associated carbon capture processes (through deforesta-tion for example). These undisputed (except by a small number of maverick scientists and political elites, and endorsed by a small but vocal part of the population) matters of fact are, without proper political intermedia-tion, translated into matters of concern. The latter, of course, are eminently political in nature. Yet, in the climate change debate, the political nature of the matter of concern is disavowed to the extent that the facts in them-selves are not only necessary, but apparently also sufficient for the elevation of their message, through a short-circuiting procedure, on to the terrain of the dignity of the political, where it is framed as a global humanitarian cause. The matters of concern are thereby relegated to a terrain beyond dispute, to ones that do not permit dissensus or disagreement. Scientific expertise becomes the foundation and guarantee for properly constituted politics/policies.

A few months after the outbreak in 2008 of the deepest and longest crisis of capitalism since the great depression, Prince Charles, heir to the throne in the United Kingdom, uttered prophetic words, announcing the coming climatic Armageddon, if no immediate action would be initiated. We had only 100 months left to act. They have now passed and very little substantial action of the sort that Charles had in mind has been done to stem climate change. The "passage to the act" was nonetheless the intention of Charles's intervention. His statement was indeed a call to arms, driven by a deep-seated belief that something serious could and should be done. His apocalyptic framing of the environmental pickle we are in is not an unusual discursive tactic. Warnings of "dangerous climate change" and pending disaster are repeated ad nauseam by many scientists, activists, business leaders, and politicians, intended primarily to nudge behavioral change and urge action. Such narratives in fact combine an unbridled optimism in humankind's capacity to act if urgency requires it and in the scientific, technological, and organizational inventiveness of some to come up with the right mix of techno-managerial measures to deflect the arrow of time such that civilization as we know it can continue a while longer.

Until 2007 and 2008, climate change and related environmental concerns were indeed fairly high on the social and political agenda, spurred on by media reports on a flood of scientific research and galvanized by popular interest and concern. Climate conferences attracted global attention only comparable to other mega-events like the Olympics, and prominent politicians biked to work or visited the Antarctic to bear witness to the facts and effects of climate change. CO_2 and other greenhouse gases nonetheless continued their seemingly unstoppable climb, to reach the emblematic 400 ppm (parts per million) in 2016, the highest concentration of CO_2 ever measured. However, an allegedly much greater catastrophe for civilization had begun to unfold when the financial crisis exploded in 2008 and pulled the core capitalist countries into the longest and deepest recession since the Great Depression. While, unsurprisingly, the output of climate gases did fall the subsequent year as Western economies contracted (but have since again begun their inexorable climb), no effort has been spared to salvage the financialized economy from its home-grown wreckage and to mobilize unprecedented public means to put the profit-train back on the rails, albeit without much success so far. Apocalyptic imaginaries of potential social and economic disintegration saturated the landscape, urging people into not only putting their trust unreservedly in the hands of the various national and global elites but also supporting the elites' clunky and desperate attempts to save their way of life.

Despite significant differences, both catastrophic narratives share an uncanny similarity, particularly if viewed from the place of elite enunciation. While the ecological Armageddon points at a universal, potentially species-wide destruction, the economic catastrophe is a particular one related solely to the threatened reproduction of, basically, capitalist relations. Yet, the discursive mobilization of catastrophe follows broadly similar lines. Imaginaries of a dystopian future are nurtured, not in the least by various political and economic elites, to invoke the specter of the inevitable if *nothing* is done so that *something will* be done. Their performative gesture is, of course, to turn the revealed (ecological or political-economic) *endgame* into a manageable *crisis*.

While catastrophe denotes the irreversible radical transformation of the existing conditions into a spiraling decline, crisis is a contingent process that requires particular techno-managerial attention by those entitled or assigned to provide it. The notion of crisis also promises the potential to contain the crisis such that the dystopian revelation is postponed or deflected. Thus, the embrace of anticipatory catastrophic discourse serves primarily to turn the nightmare into crisis management, to assure that the situation is serious but not catastrophic. Unless you are from the cynical Left—"don't panic now, we told you that crisis would come"—or from the doomsday preachers who revel in the perverse pleasures offered by the announcement of the end, the nurturing of fear, which is invariably followed by a set of techno-managerial fixes, serves precisely to depoliticize. Nurturing fear also serves to leave the action to those who promise salvation, to insist that the Big Other does exist, and to follow the Master who admits that the situation is grave, but insists that homeland security (ecological, economic, or otherwise) is in good hands. We can safely continue shopping!

What we are witnessing is a strange reversal whereby the specter of economic or ecological catastrophe or both is mobilized primarily by the elites from the Global North. Neither Prince Charles nor Al Gore can be accused of revolutionary zeal. For them, the ecological condition is—correctly of course—understood as potentially threatening to civilization as we know it. At the same time, their image of a dystopian future functions as a fantasy that sustains a practice of adjusting things today such that civilization as we know it (neoliberal capitalism) can continue for a bit longer, spurred on by the conviction that radical change can be achieved without changing radically the contours of capitalist eco-development. The imaginary of crisis and potential collapse produces an ecology of fear, danger, and uncertainty while reassuring the population that the techno-scientific and socioeconomic elites have the necessary adaptive toolkit to readjust the machine such that things can stay basically as they are. What is of course radically disavowed in their pronouncements is the fact that many people in many places of the world already live in the socio-ecological catastrophe. The ecological Armageddon already is a reality. While the elites nurture an apocalyptic dystopia that can nonetheless be avoided (for them), the majority of the world already lives "within the collapse of civilization" (The Invisible Committee 2009). The Apocalypse is indeed a combined and uneven one, both over time and across space (Williams 2011).

A flood of literature on the relationship between apocalyptic imaginaries, popular culture, and politics has excavated the uses and abuses of revelatory visions (Skrimshire 2010; Williams 2011). Despite the important differences between the transcendental biblical use of the apocalypse and the thoroughly material and socio-physical ecological catastrophes-to-come, the latter, too, depoliticize matters. As Alain Badiou contends: "The rise of the "rights of Nature" is a contemporary form of the opium of the people. It is an only slightly camouflaged religion: the millenarian terror, concern for everything save the properly political destiny of peoples, new instruments for control of everyday life, the obsession with hygiene, the fear of death and catastrophes... It is a gigantic operation in the depoliticization of subjects" (Badiou 2008b, 139).

Environmental problems are indeed commonly staged as universally threatening to the survival of humankind, announcing the premature termination of civilization as we know it and sustained by what Mike Davis aptly called "ecologies of fear" (Davis 1998). Such nurturing of fear, in turn, is sustained in part by a particular set of phantasmagorical imaginaries that serve to reinforce the seriousness of the situation (Katz 1995). The apocalyptic vision of a world without water or at least with endemic water shortages;

the ravages of hurricanes whose intensity is amplified by climate change; pictures of scorched land as global warming shifts the geo-pluvial regime and the spatial variability of droughts and floods; icebergs that disintegrate; alarming reductions in biodiversity as species disappear or are threatened by extinction; post-apocalyptic images of nuclear wastelands; the threat of peak-oil; the devastations raked by wildfires, tsunamis, spreading diseases like SARS, avian flu, Ebola, superbugs, or HIV—all these imaginaries of a Nature out of sync, destabilized, threatening, and out of control are paralleled by equally disturbing images of a society that continues piling up waste, pumping CO_2 into the atmosphere, recombining DNA, deforesting the earth, and so on. In sum, our ecological predicament is sutured by millennialism fears sustained by an apocalyptic rhetoric and representational tactics, and by a series of performative gestures signaling an overwhelming, mind-boggling danger—one that threatens to undermine the very coordinates of our everyday lives and routines and may shake up the foundations of all we took and take for granted.

Of course, apocalyptic imaginaries have been around for a long time as an integral part of Western thought, first of Christianity and later emerging as the underbelly of fast-forwarding technological modernization and its associated doomsday thinkers. However, present-day millennialism preaches an apocalypse without the promise of redemption. Saint John's biblical apocalypse, for example, found its redemption in God's infinite love, while relegating the outcasts to an afterlife of permanent suffering. The proliferation of modern apocalyptic imaginaries also held up the promise of redemption: the horsemen of the apocalypse, whether riding under the name of the proletarian, technology, or capitalism, could be tamed with appropriate political and social revolutions. The environmental apocalypse, in contrast, takes different forms. It is not immediate and total (but slow and painful), not revelatory (it does not announce the dawn of a new rose-tinted era), no redemption is promised (for the righteous ones), and there are no outcasts. Indeed, when the ship goes done, the first-class passengers will also drown.

As Martin Jay argued, while traditional apocalyptic versions still held out the hope for redemption, for a "second coming," for the promise of a "new dawn," environmental apocalyptic imaginaries are "leaving behind any hope of rebirth or renewal … in favor of an unquenchable fascination with being on the verge of an end that never comes" (Jay 1994, 33).

The emergence of new forms of millennialism around the environmental nexus is indeed of a particular kind that promises neither redemption nor realization. As Klaus Scherpe insists, this is not simply apocalypse now, but apocalypse forever. It is a vision that does not suggest, prefigure, or expect the necessity of an event that will alter the course of history (Scherpe 1987). Derrida (referring to the nuclear threat in the 1980s) sums this up most succinctly: "here, precisely, is announced—as promise or as threat—an apocalypse without apocalypse, an apocalypse without vision, without truth, without revelation ... without message and without destination, without sender and without decidable addressee ... an apocalypse beyond good and evil" (Derrida 1982). The environmentally apocalyptic future, forever postponed, neither promises redemption nor does it possess a name, a positive designation.

The attractions of such an apocalyptic imaginary are related to a series of characteristics. In contrast to standard left arguments about the apocalyptic dynamics of unbridled capitalism, I would argue that sustaining and nurturing apocalyptic imageries are an integral and vital part of the new cultural politics of capitalism for which the management of fear is a central leitmotiv (Badiou 2008c) and provide part of the cultural support for a process of post-politicization. At the symbolic level, apocalyptic imaginaries are extraordinarily powerful in disavowing or displacing social conflict and antagonisms. Apocalyptic imaginations are decidedly populist and foreclose a proper political framing. Or in other words, the presentation of climate change as a global humanitarian cause produces a thoroughly depoliticized imaginary, one that does not revolve around choosing one trajectory (such as neoliberalism) rather than another (such as eco-socialism, for example) or identifying clear adversaries in a political process; it is an imaginary that is not articulated with specific political programs or socio-ecological projects or transformations. It insists that we have to make sure that radical techno-managerial and socio-cultural transformations, organized within the horizons of a capitalist order that is beyond dispute, are initiated that retrofit the climate. In other words, we have to change radically, but within the contours of the existing state of the situation—"the partition of the sensible" in Jacques Rancière's words, so that nothing really has to change.

The negative desire for an apocalypse that few really believe will really happen (if we truly believed what we're told, that the earth is in a state of near-collapse, we would not be sitting around writing and reading arcane academic books) finds its positive injunction around a fetishist invocation of CO_2 as the "thing" around which our environmental dreams, aspirations, and contestations as well as policies crystallize. The *point de capiton*, the quilting point through which the signifying chain weaves a discursive matrix of meaning and content for the climate change problematic, is CO_2—the *objet petit a* that expresses our deepest fears and around which the desire for change, for a better socio-climatic world, circulates.

The fetishist disavowal of the multiple and complex relations through which environmental changes unfold finds its completion in the double reductionism to this singular socio-chemical component (CO_2). The reification of complex processes to a thing-like object-cause in the form of a socio-chemical compound around which our environmental desires crystallize is further inscribed with a particular social meaning and function through its enrolment as commodity in the processes of capital circulation and market exchange (Bumpus and Liverman 2008; Liverman 2009). The commodification of CO_2—primarily via the Kyoto protocol and various offsetting schemes—in turn, has triggered a rapidly growing derivatives market of futures and options. On the European climate exchange, for example, trade in CO_2 futures and options grew from zero in 2005 to 463 million tons in June 2009, with prices fluctuating from over €30 to less than €10 per ton over this time period (see www.ecx.eu, accessed July 30, 2009). After the crisis hit and combined with an oversupply of assigned carbon credits, the price of CO_2 collapsed in most markets and most trading systems found themselves in serious difficulties. There is indeed an uncanny relationship between the financialization of everything under neoliberal capitalism and the managerial and institutional architecture of carbon-trading schemes.

The proposed transformations often take a distinct dystopian turn when the Malthusian specter of overpopulation is fused with concerns with the climate, whereby, perversely, newborns are identified as the main culprits of galloping climate change and resource depletion, a view supported by luminaries like Sir David Attenborough (OM, CH, CVO, CBE), Dr. Jane Goodall (DBE), Dr. James Lovelock (CBE), and Sir Crispin Tickell (GCMG, KCVO), among others (see www.optimumpopulation.org, accessed August 2, 2010). Eco-warrior and Gaia theorist James Lovelock put it even more chillingly: "What if at some time in the next few years we realize, as we did in the 1940s, that democracy had temporarily to be suspended and we had to accept a disciplined regime that saw the U.K. as a legitimate but limited safe haven for civilization. ... Orderly survival requires an unusual degree of human understanding and leadership and may require, as in war, the suspension of democratic government for the duration of the survival emergency" (Lovelock 2010, 94–95).

Of course, the economy is greening, sustainable policies and practices are now part of the standard toolkit of any private or public actor, carbon is traded, trees are planted, activists act, energy efficiency increases, and technologies are retrofitted to produce an adaptive and resilient socio-ecological environment. Nonetheless, greenhouse gases keep on rising, and old and new fossil energy sources continue to be exploited (coal, fracking, and tar sands in particular). "Greening" the economy does not seem to deflect the process of disastrous socio-environmental transformation. Consider, for example, how the push for electrical vehicles in the name of ecological sustainability will increase the demand for electricity generation to levels that no alternative energy system can hope to meet. In the meantime, climate change scientists continue to crunch their numbers and calibrate their models. There is growing consensus now that the internationally agreed objective to keep global warming below 2° Celsius can no longer be achieved, regardless of measures taken. Even a global temperature rise of 4° Celsius seems inevitable now, while some fear, if things continue as they are, that even greater temperature increases are very likely (New et al. 2011). A four-degree rise will have profound effects and, in all likelihood, will push climate behavior over the tipping point whereby catastrophic change is inevitable, unleashing very fast and unpredictable geographical transformations in climate patterns. It is a bleak picture, one that will undoubtedly dwarf the already doomsday-laden imaginary that Prince Charles painted.

The preceding summary of the uses of apocalyptic imaginaries, the science-politics short-circuiting, and the privatization of the climate through the commodification of CO_2 is strictly parallel, I contend, with the deepening consolidation of a political populism that characterizes the present process of post-politicization. And that is what we shall turn to next.

Succumbing to the Populist Temptation

Environmental politics and debates over sustainable futures in the face of pending environmental catastrophe signal a range of populist maneuvers that nurture the consolidation of the process of post-politicization. In this section, we shall chart the characteristics of populism (see, among others, Canovan 1999, 2005; Laclau 2005; Mudde 2004; Žižek 2006a) as they are expressed in mainstream climate concerns. In other words, to the extent that consensual climate change imaginaries, arguments, and policies reflect processes of depoliticization, this consensus is sustained by a series of decidedly populist gestures. Here, I shall summarize the particular ways in which climate change expresses some of the classic tenets of populism.

First, the climate change conundrum is not only portrayed as global, but is constituted as a universal humanitarian threat. We are all potential victims. *The* environment and *the* people, humanity as a whole in a material and philosophical manner, are invoked and called into being as the objects of concern. Humanity (as well as large parts of the nonhuman world) is under threat from climatic catastrophes. However, the people are not constituted as heterogeneous political subjects, but as universal victims, suffering from processes beyond their control. The people as a political category is transformed into a generic population subject to immunological biopolitical governance. As such, populism cuts across the idiosyncrasies of different and often antagonistic human and nonhuman natures and their specific acting outs, and silences ideological and other constitutive social differences and disavows conflicts of interests by distilling a common threat or challenge to both Nature and Humanity. As Žižek puts it "populism occurs when a series of particular 'democratic' demands [in this case, a good environment, a retrofitted climate, a series of socio-environmentally mitigating actions] is enchained in a series of equivalences, and this enchainment produces 'people' as the universal political subject, ... and all different particular struggles and antagonisms appear as part of a global antagonistic struggle between 'us' (people) and 'them'" [in this case 'it,' i.e. CO_2] (Žižek 2006a, 553).

Second, this universalizing claim of the pending catastrophe is socially homogenizing. Although geographical and social differences in terms of the effects of environmental change are clearly recognized and detailed, these differences are generally mobilized to further reinforce the global threat that faces the whole of humankind (see Hulme 2008). It is this sort of argumentation that led the reports of the Intergovernmental Panel on Climate Change (IPCC) to infer that the poor will be hit first and hardest by climate change (Intergovernmental Panel for Climate Change 2009), which is of course a correct assertion—the poor are by definition ill-equipped to deal with any sort of change beyond their control—but the report continues that, therefore, in the name of the poor, climate change has to be tackled urgently.

A third characteristic of environmental populist-apocalyptic thought is that it reinforces the nature-society dichotomy and the causal power of nature to derail civilizations. It is this process that Neil Smith refers to as "nature washing": "Nature-washing is a process by which social transformations of nature are well enough acknowledged, but in which that socially changed nature becomes a new super determinant of our social fate. It might well be society's fault for changing nature, but it is the consequent power of that nature that brings on the apocalypse. The causal power of nature is not compromised but would seem to be augmented by social injections into that nature" (Smith 2008a, 245).

While the part-anthropogenic process of the accumulation of greenhouse gases is readily acknowledged, the related ecological problems are externalized, as are the solutions. CO_2 becomes the fetishized stand-in for the totality of the climate change calamities and, therefore, it suffices to reverse atmospheric CO_2 build-up to a negotiated idealized point in history, to return to climatic status quo ex ante. An extraordinary techno-managerial apparatus is under way, ranging from new eco-technologies and geo-engineering systems of a variety of kinds to unruly complex managerial and institutional configurations, with a view to producing a socio-ecological fix to make sure nothing really changes. Stabilizing the climate seems to be a condition for capitalist life as we know it to continue. Moreover, the mobilized mechanisms to arrive at this allegedly more benign (past) condition are actually those that produced the problem in the first place (the galloping commodification of all manners of nature), thereby radically disavowing the social relations and processes through which this hybrid socio-natural quasi-object came into its problematic being. Populist discourse "displaces social antagonism and constructs the enemy. In populism, the enemy is externalized or reified into a positive ontological entity [excessive CO_2] (even if this entity is spectral) whose annihilation would restore balance and justice" (Žižek 2006a, 555). The enemy is always externalized and objectified. Populism's fundamental fantasy is that of an "Intruder," or more usually a group of intruders, who have *corrupted* the system. For xenophobic politics, the refugee, the Islamist, or the Jew customarily stand in for such imaginaries of intruders that undermine the cohesion of the social fabric. In the context of climate change, CO_2 and other green-house gases stand as the classic example of a fetishized and externalized foe that requires dealing with if sustainable climate futures and cohesive socio-ecological configurations are to be attained. Problems therefore are not the result of the "system," of unevenly distributed power relations, of the class dynamics of capitalism, of the networks of control and influence, of rampant injustices, or of a fatal flow inscribed in the system, but are blamed on an outsider (Žižek 2006a, 555). That is why the solution can be found in dealing with the pathological phenomenon, the resolution for which resides in the system itself. It is not the system that is the problem, but its pathological syndrome that is posited as excess to be dealt with. While CO_2 is externalized as the socio-climatic enemy, a potential cure in the guise of the Kyoto principles is generated from within the market functioning of the system itself. The enemy is, therefore, always vague, ambiguous, socially empty or vacuous, and homogenized (like CO_2); the enemy is a mere thing, not socially embodied, named, and counted. While a proper analysis and politics would endorse the view that CO_2-as-crisis stands as the pathological symptom of the normal, one that expresses the excesses inscribed in the very normal functioning of the system (i.e., capitalism), the dominant policy architecture around climate change insists that this excessive state is not inscribed in the normal functioning of the system itself, but an aberration that can be healed by mobilizing the very inner dynamics and logic of the system (privatization of CO_2, commodification, and liberal market exchange via carbon and carbon-offset trading). In other words, an immunological prophylactic can be devised that will keep the excess in check, so that things can go on as before.

Fourth, populism is based on a politics of "the people know best" (although the latter category remains often empty, unnamed), supported by an assumedly neutral scientific technocracy, and advocates a direct relationship between people and political participation. It is presumed that enhanced participation will lead to a good, if not optimal, and democratically supported solution. This is a view strangely at odds with the presumed radical openness, uncertainty, and undecidability of the excessive risks associated with Ulrich Beck's or Anthony Giddens's second modernity. The architecture of populist governing takes the form of stakeholder participation or forms of participatory governance that operate—as argued in chapter 1—beyond-the-state and permits a form of self-management, self-organization, and controlled self-disciplining within a broadly market-led political-economic configuration.

Fifth, populist tactics do not identify a privileged subject of change (like the proletariat for Marxists, women for feminists, or the creative class for competitive capitalism), but instead invoke a common condition or predicament, the need for common humanity-wide action, mutual collaboration, and cooperation. There are no internal social tensions or internal generative conflicts; the people—in this case, global humanity—are called into being as political subject, thereby disavowing the radical heterogeneity and antagonisms that cut through the people. It is exactly this disavowed constitutive split of the people, the recognition of radically differentiated if not opposing social, political, or ecological desires, that calls the democratic political into being.

Sixth, populist demands are always addressed to the elites. Populism as a project addresses biopolitical demands to the ruling elites (getting rid of immigrants, protection from risk, saving the climate, etc.); it is not about replacing the elites, but a call on the elites to undertake action. The ecological problem and its framing are indeed no exception. An environmental populism does not invite a transformation of the existing socio-ecological order but calls on the elites to undertake action such that nothing really has to change so that everyday life can basically go on as before. In this sense, environmental populism is inherently reactionary, a key ideological support structure for securing the socio-political status quo. It is inherently nonpolitical and nonpartisan. A Gramscian "passive revolution" has taken place over the past few years, whereby the elites have not only acknowledged the climate conundrum and, thereby, answered the call of the people to take the climate seriously, but are moving rapidly to convince the world that indeed, capitalism cannot only solve the climate riddle, but that capitalism also can make a new climate by unmaking the one it has coproduced over the past few hundred years through a series of extraordinary techno-natural and eco-managerial fixes. Not only do the elites take these particular demands of the people seriously, they also mobilize them in ways that strengthen the system.

Seven, no proper names are assigned to a post-politicizing populist politics (Badiou 2005b). Post-political populism is associated with a politics of not naming in the sense of giving a definite or proper name to its domain or field of action. Only vague concepts like climate change policy, biodiversity policy, or a vacuous sustainable policy replace the proper names of politics. These proper names, according to Jacques Rancière (1995b; see also Badiou 2005b) are what constitute a genuine democracy, that is a space where the unnamed, the uncounted, and, consequently, unsymbolized become named and counted. Consider, for example, how class struggle in the nineteenth and twentieth century was exactly about naming the proletariat, its counting, symbolization, narration, and consequent entry into the techno-machinery of the state. In the twentieth century, feminist politics became named through the narration, activism, and symbolization of "woman" as a political category. And for capitalism, the "creative class" is the revolutionary subject that sustains its creatively destructive transformations. Climate change has no positively embodied name or signifier; it does not call a political subject into being. In other words, the future of a globally warmer world has no proper name. In contrast to other signifiers that signal a positively embodied content with respect to the future (like socialism, communism, liberalism), an ecologically and climatologically different future world is only captured in its negativity; a pure negativity without promises of redemption, without a positive injunction that transcends/sublimates negativity in a positive dialectic of emancipation in an ecologically sane environment, and without proper subject. The realization of this apocalyptic promise is forever postponed, the neverland of tomorrow's unfulfilled and unrealizable promises. Yet, the gaze on tomorrow permits recasting social, political, and other pressing issues today as future conditions that can be rescripted retroactively as a techno-managerial issue. Poverty and ecological problems of all kinds will eventually be sorted out by dealing with CO_2 today.

The eighth and final characteristic of populism takes this absence of a positively embodied signifier further. As particular demands are expressed (get rid of immigrants, reduce CO_2) that remain particular, populism forecloses universalization as a positive socio-environmental project. In other words, the environmental problem does not posit a positive and named socio-environmental situation, an embodied vision, a desire that awaits realization, a fiction to be realized. In that sense, populist tactics do not solve problems; they are moved around. Consider, for example, the current argument over how the nuclear option is again portrayed as a possible and realistic option to secure a sustainable energy future and as an alternative to deal both with CO_2 emissions and peak-oil. The redemption of our CO_2

quagmire is found in replacing the socio-ecologically excessive presence of CO_2 with another socio-natural object, U235/238, and the inevitable production of all manner of socio-natural transuranic elements. The nuclear fix is now increasingly staged as one of the possible remedies to save both climate and capital. It hardly arouses expectations for a better and ecologically sound society. Something similar, of course, holds for fracking or the array of Promethean geo-engineering proposals that litter the terrain of terraforming interventions allegedly capable of mitigating or adapting to the new anthropocenic world we inhabit (Shellenberger and Nordhaus 2007; Hamilton 2013).

We are now in a position to situate this argument within the broader debate about the changing nature of politics, particularly in the global North, the tactics and processes of depoliticization and the emergence of a post-political and post-democratic frame. This is what we shall turn to next.

Rethinking the Political Environment

Against thoughts of the end and catastrophe, I believe it is possible and necessary to oppose a thought of political precariousness. (Rancière 2004, 8).

In this chapter, I have argued thus far that the particular framing of climate change and its associated populist politics forecloses or disavows (or at least attempts to do so) politicization and evacuates dissent through the formation of a particular regime of environmental governance that revolves around consensus, agreement, participatory negotiation of different interests, and technocratic expert management in the context of a nondisputed management of market-based socioeconomic organization. Even a cursory analysis of green politics, whether from the perspective of environmental movements (like Greenpeace) or environmental parties (the German Greens are a classic case), over the past few decades would signal their rapid transformation from engaging in a politics of contestation, organized acting, radical disagreement, and developing visionary alternatives to their integration into stakeholder-based arrangements aimed at delivering a negotiated policy articulated around particular technical and institutional architectures.

A consensual post-politics emerges here, one that either eliminates fundamental conflict or elevates it to antithetical ultra-politics. The consensual times we are currently living in have thus eliminated a genuine political space of disagreement. These post-political climate change policies rest on the following foundations. First, the social and ecological problems caused by modernity/capitalism are external side effects; they are not an inherent and integral part of the relations of liberal politics, class power, and capitalist economies. Second, a strictly populist politics emerges here; one that elevates the interest of an imaginary "the People," Nature, or "the environment" to the level of the universal rather than opening spaces that permit us to universalize the claims of particular socio-natures, environments, or social groups or classes. Third, these side effects are constituted as global, universal, and threatening. Fourth, the "enemy" or the target of concern is continuously externalized and socially disembodied, is always vague, ambiguous, unnamed and uncounted, and ultimately empty. Fifth, the target of concern can be managed through a consensual dialogical politics whereby demands become depoliticized and politics naturalized within a given socio-ecological order for which there is ostensibly no real alternative.

The post-politicizing environmental consensus, therefore, is one that is radically reactionary, one that forestalls the articulation of divergent, conflicting, and alternative trajectories of future socio-environmental possibilities and of human-human and human-nature articulations and assemblages. It holds on to an engineering or geo-constructionist view of nature while reproducing, if not solidifying, a liberal-capitalist order for which there seems to be no alternative. Much of the climate change argument has evacuated the politics of the possible, the radical contestation of alternative future socio-environmental possibilities and socio-natural arrangements, and silences the antagonisms and conflicts that are constitutive of our socio-natural orders by externalizing conflict. It is inherently reactionary. As Alain Badiou argues, "proper" politics must revolve around the construction of great new fictions that create real possibilities for constructing different socio-environmental futures (Badiou 2005b). To the extent that the current post-political condition that combines apocalyptic environmental visions with a hegemonic neoliberal view of social ordering constitutes one particular fiction, there is an urgent need for different stories and fictions that can be mobilized for realization. This requires foregrounding and naming different socio-environmental futures and recognizing conflict, difference, and struggle over the naming and trajectories of these futures. Socio-environmental conflict, therefore, should not be subsumed under the homogenizing mantle of a populist environmentalist-sustainability discourse, but should be legitimized as constitutive of a democratic order. This, of course, turns the climate question into a question of democracy and its meaning. It asserts the horizon of a recuperated democracy as the terrain for expressing conflict, for nurturing agonistic debate and disagreement, and, most important, for the naming of different possible socio-environmental futures.

In light of the preceding, what would be possible progressive responses? I discern broadly three perspectives. The first one centers on nudging behavioral change in a more sustainable direction. Under the mantra of "it is better to do something (like recycling, growing organic food, and the like) rather than nothing," many liberal environmentalists mobilize the apocalyptic imaginary in an effort to encourage individuals to modify attitudes and behavior, and to impress on politicians and business leaders the need to heed the environmental clarion call. Sustainability hinges here on individual preference and consumer sovereignty. They also insist that the catastrophe can still be averted if proper action is taken, action that does not necessarily overhaul social relations but does postpone the environmental catastrophe so that life as we know it can continue for a while longer.

The second strand fully endorses the environmental cataclysm and revels in the certainty that this had already been predicted a long time ago. The standard response here is, "you see, we told you so." This is strictly parallel to forms of Marxist analysis of the current financial crisis in capitalism: "Don't complain now, we did tell you so." Against this cynical stand, the third, and for me proper radical response, to the apocalyptic imaginary is twofold and cuts through the deadlock embodied by the first two responses. To begin with, the revelatory promise of the apocalyptic narrative has to be fully rejected. In the face of the cataclysmic imaginaries mobilized to assure that the apocalypse will *not* happen (if the right techno-managerial actions are taken), the only reasonable response is "Don't worry (Al Gore, Prince Charles, many environmental activists, etc.), you are right but you do not go far enough. The environmental apocalypse *will* not only happen, it has already happened, *it is already here*." Many are already living in the post-apocalyptic interstices of life, whereby the fusion of environmental transformation and precarious social conditions renders live "bare." The fact that the socio-environmental imbroglio has already passed the point of no return has to be fully asserted. The socio-environmental Armageddon

is under way for many; it is not some distant dystopian promise mobilized
to trigger response today. Water conflicts, struggles for food, environmental
refugees, and more, testify to the very real socio-ecological predicament
that choreographs everyday life for a large segment of the world's population. Things are already too late; they have always already been too late.
There is no Arcadian place, time, or environment to return to, no benign
socio-ecological past that needs to be recuperated, maintained, or stabi-
lized. Many now live in the interstices of the apocalypse, albeit a combined
and uneven one. It is only within the realization of the apocalyptic reality
of the now that a new politics might emerge.

The second gesture is to reverse the order between the universal and
the particular that today dominates the catastrophic political imaginary.
This order maintains that salvaging the particular historical-geographical
configuration we are in depends on rethinking and reframing the human-
environment articulation in a universal sense. We have to change our relationship with nature, so that capitalism can continue somehow. Not only
does this argument to preserve capitalism guarantee the prolongation of
the combined and uneven apocalypse of the present, it forecloses considering fundamental change to the actually existing unequal forms of organizing the society-environment relations. Indeed, the apocalyptic imaginary is
one that generally still holds on to a dualistic view of nature and culture.
The argument is built on the view that humans have perturbed the ecological dynamic balance in ways inimical to human (and possibly nonhuman) long-term survival, and the solution consists broadly in bringing
humans (in a universal sense) back in line with the possibilities and constraints imposed by ecological limits and the nonlinear dynamics of natural
systems. A universal transformation is required in order to maintain the
present. And this can and should be done through managing the present
particular configuration. This is the message of Al Gore and Prince Charles
and many other environmental pundits. A progressive socio-environmental perspective, in contrast, has to insist that we need to transform this
universal message into a particular one. The historically and geographically
specific dynamics of capitalism have banned an external nature radically
to a sphere beyond earth. On earth, there is no external nature left. It is
from this particular historical-geographical configuration that a radical politics of transformation must be thought and practiced. Only through the

transformation of the particular socio-ecological relations of capitalism can a generic egalitarian, free, and common reordering of the human/nonhuman imbroglios be forged.

Those who already recognized the irreversible dynamics of the socio-environmental conundrum that has been forged over the past few centuries coined a new term to classify the epoch we are in. "Welcome to the Anthropocene" became a popular catch-phrase to inform us that we are now in a new geological era, one in which humans are coproducers of the deep geological time that hitherto had slowly grinded away irrespective of humans' dabbling with the surface layers of earth, oceans, and atmosphere. Nobel Prize-winning chemist Paul Crutzen introduced "the Anthropocene," coined about two decades ago as the successor name of the Holocene, the relatively benign geo-climatic period that allegedly permitted agriculture to flourish, cities to be formed, and humans to thrive (Crutzen and Stoermer 2000). Since the beginning of industrialization, so the Anthropocenic argument goes, humans' increasing interactions with their physical conditions of existence have resulted in a qualitative shift in geo-climatic acting of the earth system. The Anthropocene is nothing other than the geological name for capitalism *with* nature. Acidification of oceans, biodiversity transformations, gene displacements and recombinations, climate change, big infrastructures effecting the earth's geodetic dynamics, among others, resulted in knotting together "natural" and "social" processes such that humans have become active agents in co-shaping earth's deep geological time. Now that the era has been named the Anthropocene, we can argue at length over its meaning, content, existence, and modes of engagement (see Castree 2014, 2015; Swyngedouw and Ernstson 2018). Nonetheless, it affirms that humans and nature are coproduced and that the particular historical epoch that goes under the name of capitalism forged this mutual determination.

The Anthropocene is just another name for insisting on Nature's death. This cannot be unmade, however hard we try. The past is forever closed and the future—including nonhuman future—is radically open, up for grabs. Indeed, the affirmation of the historical-geographical co-production of society *with* nature radically politicizes nature, makes all kind of natures enter into the domain of contested socio-physical relations and assemblages. We cannot escape "producing nature"; rather, it forces us to make choices about what socio-natural relations we wish to nurture and what terraforming worlds we wish to inhabit. It is from this particular position, therefore, that the environmental conundrum ought to be approached so that a qualitative transformation of *both* society *and* nature has to be envisaged.

This perspective moves the gaze from thinking through a "politics of the environment" to "politicizing the environment" (Swyngedouw 2014). The human world is now an active agent in shaping the nonhuman world. This extends the terrain of the political to domains hitherto left to the mechanics of nature. The nonhuman world becomes enrolled in a process of politicization. And that is precisely what needs to be fully endorsed. The Anthropocene opens up a terrain whereby different natures can be contemplated and actually coproduced. And the struggle over these trajectories and, from a leftist perspective, the process of the egalitarian socio-ecological production of the commons of life is precisely what our politics are all about. Yes, the apocalypse is already here, but do not despair, let us fully endorse the emancipatory possibilities of apocalyptic life.

Perhaps we should modify the now over-worked statement of the Italian Marxist Amadeo Borigio that "if the ship goes down, the first-class passengers drown too." Amadeo was plainly wrong. Remember the movie *Titanic* (as well as the real catastrophe). A large number of the first-class passengers found a lifeboat; the others were trapped in the belly of the beast. Indeed, the social and ecological catastrophe we are already in is not shared equally. While the elites fear both economic and ecological collapse, the consequences and implications are highly uneven. The elite's fears are indeed only matched by the actually existing socio-ecological and economic catastrophes many already live in. The apocalypse is combined and uneven. And it is within this reality that political choices have to be made and sides taken. It is precisely the politicization of the environment that will be the key theme of chapters 6 and 7.

6 Urbanization and Environmental Futures: Politicizing Urban Political Ecologies

Urbanism is the mode of appropriation of the natural and human environment by capitalism. (Debord 1994, 121)

Sometime in 2013, the earth passed the symbolic threshold of 400 parts per million of CO_2 in the atmosphere. In the same year, the Fifth Assessment Report of the IPCC concluded that "most aspects of climate change will persist for many centuries even if emissions of CO_2 are stopped" (Intergovernmental Panel for Climate Change 2013, 27). Despite the migrating circuses of the UN's Climate Summits and their dismal record of broken or underperforming commitments, preciously little has been achieved in lowering total greenhouse gas emissions. Overall greenhouse gas emissions keep increasing year on year with the exception of the post-economic crisis year 2011. In the meantime, cities in both the Global North and South are choking as the concentration of small particles and other forms of pollution reach dangerously high levels.

We have now truly entered what Paul Crutzen in 2000 tentatively named the Anthropocene (Crutzen and Stoermer 2000), the successor geological period to the Holocene, and planetary urbanization is not only its geographical form: more important, urbanization is also the socio-spatial process that shapes the intimate and accelerating fusion of social and physical transformations and metabolisms that gave the Anthropocene its name (Swyngedouw 2014, 2017b). Planetary urbanization refers to the fact that every nook and cranny of the earth is now directly or indirectly enrolled in assuring the expanding reproduction of the urbanization process. Indeed, the sustenance of actually existing urban life is responsible for 80 percent of the world's greenhouse gas emissions (Bulkeley and Betsill 2005), for the intensifying mobilization of all manner of natures, and for producing most of the world's waste. From this perspective, we are in this chapter not primarily concerned with the city as a dense and heterogeneous assemblage of accumulated socio-natural things and gathered bodies in a concentrated space, but rather with the particular forms of capitalist urbanization as a socio-spatial processes whose functioning is predicated upon ever longer, often globally structured, socio-ecological metabolic flows. These flows not only weld together things, natures, and peoples, but do so in socially, ecologically, and geographically articulated but uneven manners (Cook and Swyngedouw 2012; Angelo and Wachsmuth 2015). The key question, therefore, is not about what kinds of natures are present *in* the city, but rather about the capitalist form of urbanization *of* natures: the process through which all manner of nonhuman "stuff" is socially mobilized, discursively scripted, imagined, economically enrolled and commodified, politically contested, institutionally regulated, and physically metabolized/transformed to produce socio-ecological assemblages that support the urbanization process (Heynen, Kaika, and Swyngedouw 2006). Consider, for example, how dependent are the purportedly dematerialized affective economies that animate much of contemporary urban social and cultural life (IT networks, social media, smart networks, eco-architecture, informatics, and the like) on mobilizing a range of minerals (like coltan, short for columbite-tantalite); upon feverish resource grabbing, often through tactics of dispossession, in socio-ecologically vulnerable places; upon production chains that are shaped by deeply uneven and often dehumanizing socio-ecological metabolisms (material and immaterial production processes) in order to render it useful in IT and related hardware; and upon a recycling process that returns much of the e-waste to the socio-ecologically dystopian geographies of Mumbai's or Dhaka's suburban wastelands and its informally organized waste recyclers. Indeed, the excesses of urbanization—from (e-)waste to CO_2—are customarily decanted onto the socio-ecological dumping grounds of the global peripheries of the world's cities. The capitalist form of planetary urbanization and the socio-ecological and political-economic processes that animate its combined and uneven socio-ecological development on a world scale are now generally recognized as key drivers of anthropogenic climate change and other socio-environmental transformations such as biodiversity loss, soil erosion, large eco-infrastructures like dams, deforestation, resource extraction and deep-geological mining, pollution, and the galloping commodification of all manner of natures. Our urban fate and natures' transformations are irrevocably bound up in an intimate and intensifying metabolic—but highly uneven—patterning choreographed by the power relations that animate the reproduction of neoliberal capitalism. The configuration of this urban metabolic relationship has now been elevated to a global public concern, and a feverish search for all manner of eco-prophylactic remedies has entered the standard vocabulary of both governmental and private actors.

Indeed, the search for a smart socio-ecological urbanity that seeks out the socio-ecological qualities of eco-development, retrofitting, sustainable architecture, resilient urban governance, the commodification of environmental "services," and innovative—but fundamentally market-conforming—eco-design (Mostafavi and Doherty 2010) is often predicated upon mobilizing precarious labor and dispossessing local people from their resources and livelihoods (Caprotti 2014), while still further expanding the mobilization of the earth's resource base. As discussed in the previous chapters, this techno-managerial dispositive has now been consensually established as the frontier of architectural and planning theory, and urban design practice, presumably capable of saving both city and planet, while assuring that civilization as we know it can continue for a little longer.

Nature as the externally conditioning frame for urban life has indeed come to an end. The Anthropocenic inauguration of a socio-physical historical and thoroughly globally urbanized nature forces a profound reconsideration and rescripting of both nature and urbanization in political terms. The question is not any longer about bringing environmental issues into the domain of urban politics, but rather about how to bring the political into the urban environment. In this chapter, we shall explore first the ways in which urban thought and research have begun to incorporate political matters within urban theory and practice. Particular attention will be paid to urban environmental justice perspectives on the one hand and urban political ecology on the other. While we fully endorse the extraordinary progress that has been made in recent years, we shall insist, in the second section, that urban political-ecology needs to take the question of "the political" in "political ecology" much more seriously. A number of pointers for a politicized urban political ecology will conclude the chapter.

UrbanNatural

The urbanization process understood as a constituent part of the world's socio-ecological predicament was foregrounded in the 1970s as part of a broader concern with deteriorating environmental conditions. While the voices of eco-urban visionaries like Murray Bookchin went largely unnoticed (see White 2008), the Malthusian clarion call of pending resource depletion foreshadowed by the Club of Rome's *Limits to Growth*, raising the specter of immanent scarcity in nature, really got the global elites worried about the allegedly feeble prospects for sustaining capitalist accumulation for much longer, and pointed to urbanization as the main culprit of the world's accelerating resource depletion (Meadows et al. 1972). In addition, the environmental movement, particularly active around contesting nuclear energy use in the Global North, and hyper-urbanization in the Global South, propelled environmental matters to the top of the urban policy agenda.

Urban thought and practice followed suit. Urban scholars and activists began to dissect the urbanization of nature as a process of continuous deterritorialization and reterritorialization of socio-ecological metabolic circulatory flows, organized through predominantly capitalist social relations sustained by privately or publicly managed socio-physical conduits and networks (Swyngedouw 2006), and nurtured by particular imaginaries of what nature is or should be. Under capitalism, so the argument went, the mobilization and transformation of nonhuman stuff into a commodified form under the impetus of capital circulation and accumulation shape these socio-ecological processes and turn the city into a metabolic socio-environmental process that stretches from the immediate environment to the remotest corners of the globe. Through this conceptual lens, urbanization is viewed as a process of geographically arranged socio-environmental metabolisms that fuse the social with the physical. In so doing, a "cyborg" urbanity is produced that enmeshes distinct physical forms with cultural and socioeconomic relations in geographically highly uneven socio-ecological manners (Swyngedouw 1996a; Gandy 2005). A proliferating body of scholarly work began to explore, both empirically and theoretically, how urbanization and its human and nonhuman inhabitants across the globe are linked through socio-technological networks, flows of matter, and social relations of power for the extraction, circulation and disposal of matter such as water (Swyngedouw 2004b), energy (Bouzarovski 2104; Verdeil 2014), fat (Marvin and Medd 2006), chemicals and e-waste (Pellow 2007), food (Heynen, Kurtz, and Trauger 2012), household waste (Njeru 2006; Armiero 2014), infrastructures (Graham and Marvin 2001; Monstadt 2009), or redundant ships (Buerk 2006; Hillier 2009). Burrowing into the metabolic process of less visible, yet powerfully important socio-natural actants, Ali and Keil mapped how the SARS epidemic challenged global networks of urban governance (Ali and Keil 2008). Bulkeley searched for the urban roots of CO_2 (Bulkeley and Betsill 2005), and Robbins reconstructed the global networks of production, pollution, and toxic waste, which sustain the insatiable drive to nurture the green lawns that feed the American suburban middle-class dream (Robbins 2007).

A series of exciting urban monographs explored the political-ecological dynamics that undergirded the historical-geographical production of particular cities. For example, William Cronon's seminal monograph *Nature's Metropolis* rewrites Chicago's urbanization process through an examination of how wheat and hog production shaped the city's metabolic and spatially expanding transformation process and linked Chicago to its hinterland, to other cities, and to the global market (Cronon 1991). Brechin narrates how San Francisco's elites rummaged through nature in search of earthly gain and power (Brechin 2001) while Matthew Gandy's *Concrete and Clay* undertakes the archeology of New York's urbanization process as a political-ecological construct (Gandy 2002). In some studies, water became an emblematic entry into the excavation of socio-ecological flows. Maria Kaika's *City of Flows* considers the cultural, socioeconomic, and political relations though which the circulation of water in Athens and other cities is cast and recast during modernity (Kaika 2005). Swyngedouw's *Social Power and the Urbanization of Nature* excavates the relationship between planetary urbanization and nature's transformation through the lens of Guayaquil's water (Swyngedouw 2004b), while Karen Bakker follows the flow of water through the privatization politics of England and Wales (Bakker 2003), and David Saurí and colleagues (Masjuan et al. 2008; March and Saurí 2013) explore the political-ecological dynamics, conflicts, and struggles around Barcelona's urban water supply. It is also worth pointing to how Mike Davis's dystopian *Dead Cities and Other Tales* excavates the peculiar ecologies of cities that should not be where they are (Davis 2002), and to Freidberg's majestic study of how green beans link African cities to Paris and

London, thereby exploring how urbanization is indeed sustained by planetary socio-ecological networks and relations driven by particular dynamics of urbanization and city life (Freidberg 2004).

All of the preceding narratives in urban political ecology and cognate research demonstrated, in a variety of ways and from a range of theoretical perspectives, how the matter of matter becomes an active moment in the political-ecological transformations that shape planetary urbanization. These authors have argued convincingly that the urban process has to be theorized, understood, and managed as a socio-natural process that goes beyond the techno-managerial mediation of urban socio-ecological relations. By doing so, they have contributed to delegitimizing dominant twentieth-century perspectives on the city that ignored nature, without falling into the deadlock of nature fetishism or ecological determinism. Moreover, by transcending the binary division between nature and society, the urban metabolism perspective has shown that socio-ecological processes are intensely political and politicized, and confirmed that urban theory without nature cannot be but incomplete.

However, this body of thought has paid relatively little attention to the political possibilities such renatured understandings of urbanization could bring, or to imagining radically different possible futures for urban socio-ecological assemblages that are an integral part of radically politicizing mobilizations. Thus, although we may now be able to trace, chart, follow, and narrate the multiple socio-ecological lines and flows that shape the globalizing urban process, preciously little has been said about how to produce alternative, more equitable and enabling, urban socio-ecological constellations. In what follows we shall briefly explore two perspectives that have galvanized more politicized thinking and practice around the urban environmental question.

Urban Environmental Justice (UEJ): The Distribution of Environmental Bads

The urban environmental justice perspective opens up a politicizing view in which the unequal distributional characteristics of urban metabolism take center stage. UEJ is sensitive to the conflicting and power-laden processes of urbanizing nature through elite-led techno-managerial fixes (Walker 2012). Originating in the United States, urban environmental justice emerged both as a normative concept and a social movement, sustained by new insights into the highly uneven distribution of environmental "goods" and "bads" *in* the city. Early work in the 1980s had already begun to recognize that poor, often predominantly African-American neighborhoods were overwhelmingly located in areas characterized by environmentally hazardous conditions (Bullard 1990). Significant positive correlations were found between the presence of toxic dumps, waste processing facilities, ground pollution, hazardous chemicals, or absence of green zones on the one hand, and concentrations of low-income households on the other. In other words, the spatial distribution of environmental goods and bads mirrored the socio-spatial distribution of political power, wealth, and income (Schlosberg 2007).

Urban environmental justice became defined and understood as a question of Rawlsian distributional (in)justice, choreographed and structured by the highly uneven political and economic power relations through which decisions over environmental distributional conditions are made and implemented. Emphasis is put on the socially and economically uneven positions, modes of recognition, and capabilities of different urban dwellers in the urban political and economic decision-making machinery that allocates the distribution of environmental goods and bads throughout the city, showing that the partitioning of environmental goods mainly benefit urban elites, while environmental bads are decamped to areas where the powerless and disenfranchised live. The partitioning of the sensible in the police order, as Jacques Rancière would call it, manifests itself, among others, in unequal socio-ecological configurations with highly stratified relations of access to environmental goods or exposure to environmental bads. It became clear that sustainable urban lives are primarily the privilege of the rich, and that environmental havens are sustained on the back of deteriorating socio-ecological conditions elsewhere. More recent explorations of urban environmental injustices have extended the earlier focus on race to other social categories such as gender, class, age, ability, and geographical scale (Walker 2009). Nonetheless, the emphasis of UEJ remains clearly on foregrounding liberal notions of procedural and distributional justice as fairness, and expressing a distinct form of NIMBY-ism (Not In My Back Yard).

UEJ is primarily concerned with the procedures through which environmental technologies, infrastructures, and amenities are partitioned and distributed throughout the city and highlights the socially highly uneven patterning of ecological qualities and hazards. This perspective succeeded in socializing nature and ecology by excavating the intricate mechanisms through which nature, ecological processes, and socio-environmental conditions in the city are highly interwoven in deeply unjust manners. It demonstrates how the partitioning of socio-environmental things and processes through formal institutional and informal socio-political practices is thoroughly nonegalitarian, opening up the possibility for political interruption in the Rancièrian sense. The uneven distribution becomes etched in the urban landscape through a combination of highly elitist decision-making procedures on the one hand and their cementing into the architecture of eco-technological infrastructures and technologies on the other. Nonetheless, UEJ tends to be symptomatically silent about the particular ways in which political forms of power interweave with the concrete modalities through which nature becomes enrolled in processes of capital circulation and accumulation. Its place-based focus too needs to be augmented by considering the global metabolic assemblages and flows that produce combined and uneven socio-ecological change.

Urban Political Ecology (UPE): Reasserting the Capitalist Production of Planetary Urbanization

While UEJ focuses primarily on patterns of socio-ecological injustice within the city, urban political ecology shifts the interpretative gaze to the socio-ecological inequalities embodied in and shaped by the production and reproduction of capitalist urbanization itself (Keil 2003, 2005). Under capitalism, a wide range of different natures becomes increasingly enrolled in the circuits of capital accumulation through which they are both transformed and de-/reterritorialized. This is a socio-metabolic process whereby "physical matter such as water or cows is transformed into useable, ownable, and tradable commodities" (Coe, Kelly, and Yeung 2007, 161). From this perspective, Nature as the homogenized collective name for all manner of nonhuman things, organisms, and processes does not exist, but rather there is a highly diverse and continuously changing collection of all sorts of nonhuman imbroglios that become historically and geographically produced in specific and decidedly urbanized manners (see previous chapters). It is such a conceptualization that led David Harvey, for example, to argue that "there is nothing unnatural about New York City" (Harvey 1993, 28; 1996, 186).

UPE is decidedly anti-Malthusian. In contrast to the doom-laden demographic specter of Malthusian limits to the earth's resource base and the menace of pending absolute scarcity, urban political ecology considers scarcity as socio-ecologically produced through the twin imperative of "accumulation for accumulation's sake" on the one hand and market forces as naturalized and privileged instruments for the social allocation and distribution of (transformed) natures on the other. Furthermore, UPE rejects the apocalyptic imaginary that customarily accompanies attempts at politically foregrounding a public concern with nature as inherently depoliticizing and reactionary (Swyngedouw 2013a, 2013b). For UPE, the socio-ecological catastrophe is already present and reflects the combined and uneven socio-ecological patterns produced by the specifically capitalist form of globalization of urban metabolisms.

Indeed, a political-economic configuration—usually called capitalism—whose sustainability is predicated upon growth for growth's sake necessarily hits the physical and social limits of its own preconditions of existence, thereby ushering in continuous and highly uneven dynamics of pervasive socio-ecological restructuring and transformation. More important, such produced urban socio-physical environments embody and reflect the unequal power and associated asymmetrical socio-ecological living conditions inscribed in socio-ecological metabolisms. "Scarcity" or "socio-ecological disintegration" resides, therefore, not in Nature but in the socially constructed and utterly contingent modalities of its spatially and socio-ecologically variegated enrolling within urbanizing circuits of capital circulation and accumulation. The production of urban environments, and the metabolic vehicles (such as infrastructures of all kinds, the technical conditions that permit the flow and metabolization of, among others, energy, food, information, bodies, and things) that secure its functioning are of course mediated by institutional and governance arrangements that are often nominally democratic, but are nonetheless necessarily deeply committed to assuring the uninterrupted expansion of the capital circulation process (Virilio 1986). Metabolic vehicles are the hard and soft infrastructures through which nonhuman matter becomes transformed, and express in their techno-political functioning multiple relations of power in which social actors strive to create and defend socio-physical environments that serve their interests and satisfy their desires. It is precisely this articulation between state, class, and environmental translation that renders urban socio-ecological processes, including the question of sustainability, highly conflictive and subject to intense political and social struggle. Consider, for example, how the urban rebellion that engulfed Turkey with rarely seen intensity in the summer of 2013 emblematically sparked off with a conflict over a park and a few trees on Istanbul's Taksim Square (see also chapter 7). Or how climate summits meet with increasingly intense street protests.

The urbanization of nature is decidedly multi-scaled and spatially networked in an extended manner. Multi-scalar governance arrangements, from Agenda 21 to the Kyoto protocol, suggest how the global span of socio-ecological transformation processes are articulated with multi-scaled governance ensembles, each of which expresses particular power relations whereby struggles for control, access, and transformation of nature and the distribution of ecological goods and bads are carefully negotiated and intensely contested. From this political-ecological perspective, urban ecological conditions and the configurations of their governance are never just local, but are attached to processes that operate in diverse ecologies across the world. Such urban political-ecological approaches foreground the political core of environmental change and transformation and insist on the fundamentally political nature of the modes of socio-technically organizing the metabolic transformation of nature.

Politicizing the Political Ecology of Planetary Urbanization

Despite the extraordinary leap forward in critical understanding of the urban environmental condition and the consensual attention to sustainable and smart eco-technologies, global ecological conditions continue to deteriorate at an alarming rate as planetary urbanization intensifies. This is a veritable paradoxical situation that can only be rendered legible in strictly ideological terms. While the techno-managerial elites desperately attempt to micro- and macro-engineer socio-ecological conditions in ways that permit both sustaining economic growth indefinitely into the future and turning environmental technologies into a green accumulation strategy, the depth and extent of environmental degradation gallop further in what Calder Williams calls "a combined and uneven apocalypse" (Williams 2011).

These arguments point to the vital importance of grappling with the process of post-politicization and for moving from an urban politics of the environment to urbanizing environmental politics. If the aim of politics—including urban politics and policies, city design, planning, and architecture—is intervention that can change the given socio-environmental ordering and partitioning in a certain direction, then such intervention often constitutes a violent act, in the sense that it erases what is there (at least in part) in order to erect something new and different. The central point is to recognize that political acts are singular interventions that produce particular socio-ecological arrangements and milieus and, in doing so, foreclose the possibility of others emerging. An intervention enables the formation of certain socio-ecological assemblages while closing down others. The "violence" inscribed in such choice has to be fully acknowledged. For example, one cannot have simultaneously a truly carbon-neutral city and permit unlimited car-based mobility. They are mutually exclusive. Even less can an egalitarian, democratic, solidarity-based, and ecologically sensible urban future be produced without marginalizing or excluding those who insist on the private appropriation of the commons of the earth and its mobilization for accumulation, personal enrichment, and hereditary transfer of accumulated resources.

Such violent encounters always constitute a political act, one that can be legitimized only in political terms, and not through an externalized legitimation that resides in a fantasy of Nature or Sustainability. Any political act is one that reorders socio-ecological coordinates and patterns, reconfigures uneven socio-ecological relationships (while foreclosing others), often with unforeseen or unforeseeable consequences. Such interventions that express a choice and take sides are invariably somehow exclusive. Any instituted order, including a liberal-democratic one, produces its own inequalities. This is precisely what critical urban theory has successfully explored and analyzed over the past few decades. This suggests that there is an irreducible gap or abyss between the democratic as a political given predicated upon the presumption of the equality of each and all, on the one hand, and, on the other hand, the instituted/institutionalized forms of policy-making that invariably suspend this axiomatic equality. This gap between politics and the political needs to be endorsed fully.

Most important, it pits those who are bent on maintaining the current trajectory that produces a combined and uneven socio-ecological apocalypse radically against those who prefigure an inclusive and egalitarian production of socio-ecological urban commons. This perspective is concerned with the democratic and emancipatory political process through which such politically embedded ecological transformation takes place. Rather than invoking a normative notion of environmental justice or of an idealized (balanced) nature, a politicizing view insists on focusing on the realities of the presumed democratic political equality in the decision-making processes that organize socio-ecological transformation and choreograph the management of the commons. In doing so, the attention shifts from a techno-managerial or ethical perspective to a resolutely political vantage point—articulated around the notion of equality as explored in chapter 3—that considers the ecological conundrum to be inexorably associated with democratic political acting, and focuses on the fundamentally politicized conditions through which natures become produced.

While a pluralist democratic politics insists on difference, disagreement, radical openness, and exploring multiple possible futures, concrete spatial-ecological intervention is necessarily about relative closure (for some), definitive choice, singular intervention, and, thus, certain exclusion and occasionally even outright silencing. For example, tar sand exploitation and fracking cannot coincide with a climate policy worthy of its name. Climate justice requires a process that keeps the coal in the hole, the oil in the soil, and tar sand in the land. This would strictly outlaw dominant practices and produce extraordinary distributional effects that can only be attended to politically. While traditional democratic policies are based on majoritarian principles, the democratic-egalitarian perspective insists on foregrounding equality and socio-ecological solidarity as the foundational gesture for a green urban future.

Politicizing environments democratically, then, becomes an issue of enhancing the urban democratic political content of socio-environmental construction by means of identifying the strategies through which a more equitable distribution of social power and a more egalitarian mode of producing urban natures can be achieved. This requires the nurturing of processes that enable the production of spaces of democratization (i.e., spaces for the enunciation of agonistic dispute) as a foundation and condition for more egalitarian urban socio-ecological arrangements, and the naming of positively embodied egalibertarian socio-ecological futures that are immediately realizable. Agonism refers here to the process by which oppositional positions between enemies become articulated and organized as oppositional encounter between adversaries (Mouffe 2013). In other words, egalitarian urban ecologies are about demanding the impossible and realizing the improbable, often in the face of radical and powerful opposition, and this is exactly the challenge the Anthropocene poses. In sum, the politicization of the environment is predicated upon the recognition of the indeterminacy of nature, the constitutive split of the people, the unconditional democratic demand of political equality, and the real possibility of the inauguration of public and collective urban socio-ecological futures that express the democratic presumptions of freedom and equality.

Ultimately, the intellectual challenge posed by the socio-environmental conditions shaped by planetary urbanization must be to extend the intellectual imaginary and the powers of thought and practice to overcome the contemporary cultural impasse identified by Jameson that "it is easier to imagine the end of the world than changes in the eco-capitalist order and its inequities" (Jameson 2003, 76). This is the courage of the intellect that is now required more than ever, a courage that takes us beyond the impotent confines of a sustainability discourse and leaves the existing combined and uneven, but decidedly urbanized, socio-ecological dynamics fundamentally intact. It is a courage that charts new politicized avenues for producing a new common urbanity. There is an urgent task ahead, therefore, to delve into the complex linkages between politicizing discourse and practice, post-political eco-management, and the reproduction of environmental socio-ecological inequalities. It is necessary to ask questions about what visions of Nature and what socio-environmental relations are being promoted; what quilting points are being used and how they are being stitched together; and who are promoting these visions and why. In this respect, there is an urgent need to consider the eco-politicizing movements and discourses such as those of the environmental political movements or the various *indignados* and other insurgent political mobilizations that over the past few years have been demanding a new constituent democratic process. The articulation between urban political ecological thought and democratizing urban practices with a view toward thinking whether an ecologically sensible, equal, free, and solidarity-based form of planetary urbanization can still be imagined for the twenty-first century is, I believe, the greatest intellectual challenge for an urban political ecology that desires to be politically performative. Addressing this challenge is the theme of the final two chapters.

III Specters of the Political

In the final part of this book, attention shifts from an analytical-theoretical frame and a diagnostic of how post-politicizing modes of politics unfolded through the registers of climate change and the process of urbanization to considering the promises of the return of "the political." As discussed in part I, processes of depoliticization cannot fully suture or colonize the space of the political. The "repressed" returns in a range of processes of repoliticization that open up potential new trajectories for emancipatory change. The key problematic that animates the following chapters revolves around the question of "politicization." What precisely is being repressed, disavowed, and foreclosed in processes of post-politicization? And how and where can one discern glimpses of repoliticization in a time of post-democratic consensual governance? Our emphasis will be on thinking through real, existing or possible moments of repoliticization and how these might turn into a rescripting and reenacting of emancipatory political sequences. This is indeed the primary challenge for the twenty-first century. Either we manage the excesses of capitalism under a post-political frame to the best of our humanitarian capacities or we renew our fate not only in the possibility but also the utmost necessity of reviving the horizon of emancipation. The latter would revolve around the potential for inaugurating a mode of being together in more equal, free, and solidarity-based manners in a socio-ecologically more sensible environment.

Chapter 7 will consider the political moments inscribed in a range of urban insurgencies that have emerged since 2011 when the Arab Spring unexpectedly began to challenge the political status quo across the Middle East. Ever since, insurgent citizens from around the globe have taken to the streets, parks, and other public spaces in a collective effort to democratize the polis. The chapter discusses both the tensions and promises opened up by such insurgencies and the possibilities for initiating an emancipatory political procedure. The concluding chapter explores the wider possibilities for radical political transformation in the twenty-first century, and engages with this essential question: is it still possible to imagine and practice emancipatory geographies today?

7 Insurgent Architects, Radical Cities, and the Spectral Return of the Political

It's useless to *wait*—for a breakthrough, for the revolution, the nuclear apocalypse or a social movement. To go on waiting is madness. The catastrophe is not coming, it is here. We are already situated *within* the collapse of a civilization. It is within this reality that we must choose sides. (The Invisible Committee 2009, 138)

Insurgent Cities: Staging Equality

The Taksim Square revolt in Istanbul, the "umbrella" movement in Hong Kong, the Brazilian urban insurgencies, the French *Nuit debout,* and a wide range of other urban rebellions have been dotting the landscape of contemporary cities in recent years. Romanian activists mobilize Occupy movement tactics to fight against accumulation by dispossession, against threatened socio-environmental destruction by Canadian company Gabriel Resources' planned gold mining in Rosia Montana, and against the pending restructuring and further privatization of Romania's medical services (Velicu and Kaika 2015). The planned removal of a few trees and the construction of an Ottoman-style supermarket sparked the Taksim Square insurgency in Istanbul. These urban rebellions are just a few in a long sequence of urban political insurgencies that erupted rather unexpectedly after Mohamed Bouazizi's self-immolation on December 17, 2010, ignited the Tunisian revolution. During the magical year of 2011, a seemingly never-ending proliferation of urban insurgencies, sparked off by a variety of conditions and unfolding against the backdrop of very different historical and geographical contexts, profoundly disturbed the apparently cozy neoliberal status quo and disquieted various economic and political elites (Mayer, Thörn, and Thörn 2016). There is indeed an uncanny choreographic affinity between the eruptions of discontent in cities as diverse as Istanbul, Cairo, Tunis, Athens, Madrid,

Lyon, Lisbon, Rome, New York, Tel Aviv, Chicago, London, Berlin, Thessaloniki, Santiago, Stockholm, Hong Kong, Cape Town, Barcelona, Montreal, Oakland, Sao Paulo, Bucharest, or Paris, among many others. The *end of history* proved to be remarkably short-lived as incipient political movements staged, albeit in often inchoate, contradictory, and confusing manners, a profound discontent with the state of the situation and choreographed tentatively new urban modes of being-in-common.

A wave of deeply political protest continues to roll through the world's cities, whereby those who do not count demand a new constituent process for producing space politically. Under the generic name of *Real Democracy Now!*, these heterogeneous gatherers are outraged by and expose the variegated wrongs and spiraling inequalities of autocratic neoliberalization and existing, instituted democratic governance. The era of urban social movements as the horizon of progressive urban struggles, celebrated ever since Manuel Castells's seminal 1980s book *The City and the Grassroots* (Castells 1993), seems to be over. A much more politicized if not radical mobilization, animated by urban insurgents, is increasingly choreographing the contemporary theater of urban politicized struggle and conflict (Dikeç and Swyngedouw 2017; Swyngedouw 2017a).

It is precisely the aftermath of such urban insurrections that provides the starting point for the arguments developed in this chapter. From a progressive political perspective, the central question that has opened up, after the wave of insurgencies of the past few years petered out, centers on what to do and what to think next. Is there further thought and practice possible after the squares are cleared, the tents broken up, the energies dissipated, and everyday life resumes its routine practices? Can we discern in these movements the immanence of a new political promise?

The Spectral Return of the Political

As discussed in chapters 3 and 4, for Jacques Rancière, democratization of the polis occurs when those who do not count stage the count, perform the process of being counted, and thereby initiate a rupture in the order of things, "in the distribution of the sensible," such that things cannot go on as before (Rancière 1998). From this perspective, democratization is a performative act that both stages and defines equality, exposes a *wrong*, and aspires to transform the senses and the sensible, to render common sense what was nonsense before. Democratization, he contends, is a disruptive affair whereby the *Ochlos* (the rabble, the scum, the outcasts, "the part of no part") stages to be part of the *Demos* and, in doing so, inaugurates a new ordering of times and places, a process by which those who do not count, who do not exist as part of the polis become visible, sensible, and audible, stage the count and assert their egalitarian existence. Egalitarian politics is about "the symbolic institution of the political in the form of the power of those who are not entitled to exercise power—a rupture in the order of legitimacy and domination. It is the paradoxical power of those who do not count: the count of the 'unaccounted for'"(Rancière 2000a, 124). Egalitarian-democratic demands and practices, scandalous in the representational order of the police yet eminently realizable, are precisely those staged through mobilizations varying from the Paris and Shanghai communes to the Occupy movement, *Indignados*, and assorted other emerging political movements that express and nurture such processes of embryonic repoliticization. Identitarian positions become, in the process, transfigured into a commonality, and a new common sense, and they can be thought and practiced irrespective of any substantive social theorization—it is the political in itself at work through the process of political subjectivation, of acting in common by those who do not count, who are surplus to the police.

There are many uncounted today. Alain Badiou refers to them as the "inexistent"—the masses of the people that have no say, "decide absolutely nothing, have only a fictional voice in the matter of the decisions that decide their fate" (Badiou 2012, 56). These inexistent are the motley assortment of apolitical consumers, frustrated democrats, precarious workers, undocumented migrants, and disenfranchised citizens. The scandal of actually existing, instituted (post-)democracy in a world choreographed by oppression, exploitation, socio-ecological degradation, and extraordinary inequalities resides precisely in rendering masses of people inexistent, politically mute, without a recognized voice.

For Badiou, "a change of world is real when an inexistent of the world starts to exist in the same world with maximum intensity" (Badiou 2012, 56). In doing so, the order of the sensible is shaken and the kernel for a new common sense, a new mode of being in common becomes present in the world, makes its presence sensible and perceptible. It is the appearance of another world in the world. Was it not precisely the sprawling urban insurgencies that ignited a new sensibility about the polis as a democratic and potentially democratizing space? This appearance of the inexistent, staging the count of the uncounted is, it seems to me, what the polis, the political city, is all about. Indeed, as Foucault reminds us, "the people is those who, refusing to be the population, disrupt the system" (Foucault 2007, 43–44).

The notion of the democratizing polis introduced here is one that foregrounds intervention and rupture, and destabilizes the apparently cozy biopolitical order, sustained by an axiomatic assumption of equality. Democratization, then, is the act of the few who become the material and metaphorical stand-in for the many; they stand for the dictatorship of the democratic—direct and egalitarian—against the despotism of the instituted "democracy" of the elites—representative and inegalitarian (Badiou 2012, 59). Is it not precisely these insurgent architects that brought to the fore the irreducible distance between the democratic as the immanence of the presumption of equality and its performative spatialized staging on the one hand and democracy as an instituted form of regimented oligarchic governing on the other hand? Do the urban revolts of the past few years not foreground the abyss between "the democratic" and "democracy," the surplus and excess that escapes the suturing and depoliticizing practices of instituted governing? Is it not the reemergence of the proto-political in the urban revolts that signals an urgent need to reaffirm the urban, the polis, as a political space, and not just as a space of biopolitically governed city life?

Of course, the social markers of the insurgencies are geographically highly differentiated: the resistance against the Morsi regime in Egypt, the attacks on Erdogan's combination of religious conservatism with a booming neoliberalization of the urban process in Turkey, the accumulation by dispossession by which all sorts of people are deprived of their livelihoods, the spiraling discontent over the public bailouts and severe austerity regimes mounted by assorted states and international organizations to safeguard the global financial system (and the very socially embodied agents that sustained its growth) from immanent collapse after the speculative bubble that had nurtured unprecedented inequalities and extraordinary concentration of wealth finally burst in 2008. The quilting points that sparked these rebellions were highly variegated too: a threatened park and a few trees in Istanbul, a religious-authoritarian but nonetheless democratically elected regime in Egypt, massive austerity and housing crises in Greece, Portugal, and Spain, social and financial mayhem in the UK or the US, a rise in the price of public transport tickets in Sao Paulo, the further commodification of higher education in Montreal, large-scale gold-mining in Romania, electoral procedures in Hong Kong, changing employment law and workers' rights in France. Yet, the urban insurgents quickly turned their particular, occasionally identitarian, grievances into a wholesale attack on the instituted order, on the unbridled commodification of urban life in the interests of the few, on the highly unequal socioeconomic outcomes of actually existing representational *democracy-cum-capitalism*. The particular demands transformed quickly and seamlessly into a universalizing staging for something different, however diffuse and unarticulated this may be at present. Heterogeneous social actors came together as politicizing subjects around the signifying banner of Real Democracy Now! The assembled groups ended up without particular demands addressed to the elites, to a Master. In their refusal to express specific grievances, they demanded everything, nothing less than the transformation of the instituted order. They staged in their socio-spatial acting new ways of practicing equality and democracy, experimented with innovative and creative ways of being-together in the city, and prefigured, both in practice and in theory, new ways of distributing goods, accessing services, producing healthy environments, organizing debate, managing conflict, practicing ecologically saner lifestyles, and negotiating urban space in an emancipatory manner.

These insurgencies are decidedly urban; they may be the embryonic manifestation of the immanence of a new urban commons (see García Lamarca 2017), one that is always potentially in the making, ready to produce a new urbanity through intense meetings and encounters of a multitude, aspiring toward spatialization, that is, toward universalization. Such universalization can never be totalizing as the demarcation lines are clearly drawn, a line that separates the us (as multitude) from the them—in other words, those who mobilize all they can to make sure nothing really changes, captured neatly in the slogan of the 99 percent versus the 1 percent. The democratizing minority stands here in strict opposition to the majoritarian rule of instituted democracy. As much as the proletarian, feminist, or African-American democratizing movements were (and often still are) also very much minoritarian in terms of politically acting subjects, they nonetheless stood and stand for the enactment of the democratic presumption of equality of each and all. The space of the political disturbs the given socio-spatial ordering by rearranging it with those who stand in for "the People" or "the community" (Rancière 2001). It is a particular that stands for the whole of the community and aspires toward universalization. The rebels on Tahrir or Taksim Square are not the Egyptian or Turkish population; while being a minority, they stand materially and metaphorically for the Egyptian and Turkish people. The political emerges, Rancière attests, when the few claim the name of the many, start to embody the community as a whole, and are recognized as such. The emergence of political space is always specific, concrete, particular, and minoritarian—it is a space of appearance—but stands as the metaphorical condensation of the generic, the many, and the universal.

These attempts to produce a new commons offer perhaps a glimpse of the theoretical and practical agenda ahead. Do they not call for an urgent reconsideration of both urban theory and urban praxis? Their acting signals a clarion call to return the intellectual gaze to consider again what the polis has always been—namely, the site for political encounter. The polis is indeed the place for enacting the new, the improbable, things often considered impossible by those who do not wish to see any change. It is a site for experimentation with the staging and production of new radical imaginaries for what urban democratic being-in-common might be all about. For me, therefore, recentering the urban political is one of the central intellectual demands adequate to today's urban life.

Specters of the Urban Political Rescripted

For Alain Badiou, the political is not a reflection of something else, like the cultural, the social, or the economic. For him, the social sciences and social scientists can at best be oppositional, operating within the standard contestation of "democratic" rule (Badiou 1999, 24), but cannot conceive of the process of political transformation as the active affirmation of the egalitarian capacity of each and all to act politically. The political is a site open for occupation by those who call it into being, claim its occupation, and stage "equality" irrespective of the place they occupy within the social edifice. The political is manifested in the process of subjectivation, and appears in the "the act" of placing one's body together with others in a public space. It is precisely this process of political subjectivation that the social sciences rarely, if at all, capture (but see Velicu and Kaika 2015 and García Lamarca 2017). In what follows, I shall explore further the understanding of the political that foregrounds the notion of equality as the axiomatic

and contingent foundation of democracy. The political becomes symptomatically sensible in spatialized and collective interruptions or insurgencies
that express an unconditional egalitarian demand. It is an immanent process expressed in the rupture of any given socio-spatial order by exposing a
wrong and staging equality.

The political inaugurates a new world within the world. The insurgents'
performative and localized inscriptions become the evental time-spaces
from where a new democratizing political sequence may unfold. Insurgent
democratic politics, therefore, are radically anti-utopian; they are not about
fighting for a utopian future, but are precisely about bringing into being,
spatializing, what is already promised by the very principle upon which the
political is constituted, that is, equalitarian emancipation, and this is experimented with in the staging of democratizing spatialities. Such egalitarian
staging of being-in-common emerges and unfolds "at a distance" from the
theaters and terrains of politics and policy-making. Insurgent urbanity cannot do other than provoke the wrath of the state and has to confront, to
stare in the face, the violence that marks such potential "rebirth of history"
as Badiou provocatively calls it (Badiou 2012). Insurrectional interruption
precisely incites the objective inegalitarian violence of the instituted order
to become subjective, socially embodied, and perceptible, to render visible
the irreducible gap between the democratic as immanent process and the
police as the instituted, inegalitarian but taken-for-granted order (Žižek
2008c). Confronting the violence of the police and navigating a course
that opens up trajectories of change while preventing the confrontation to
descend into a spiraling abyss of violence is an urgent and difficult task, one
that hinges fundamentally on the process of organization and the modalities of its universalization. The political is indeed the moment of confrontation, when the principle of equality confronts a wrong instituted through
the police order. Politics understood in these terms unhinges a deep-seated
belief that expert knowledge and managerial capacity can be mobilized to
enhance the democratic governance of urban space, to limit the horizon of
intervention to consensualizing and "participatory" post-democratic management of the state of affairs (Swyngedouw 2009a, 2011). This argument
of course does not ignore or minimalize the vitally important emancipatory struggles that potentially unfold around identitarian inscriptions and/
or those struggles that are waged within the instituted order and through
which occasionally important victories are achieved in terms of enhancing

egalitarian procedures or improving the democratic content of decision-making processes within the police order.

Democratizing Spaces

Rancière's notion of the political is characterized by division, conflict, and polemic (Valentine 2005). For him, "democracy always works against the pacification of social disruption, against the management of consensus and 'stability'. ... The concern of democracy is not with the formulation of agreement or the preservation of order but with the invention of new and hitherto unauthorized modes of disaggregation, disagreement and disorder" (Hallward 2005, 34–35). The politics of consensual urban design in its post-politicizing guise, therefore, colonize the political and contribute to a further hollowing out of what, for Rancière and others, constitutes the very horizon of the political as a radically heterogeneous and conflicting one. Disavowal of the political is pushed to its limits in such process of foreclosure. Indeed and ironically, by inviting debate and discussion that eschews rupture, the political is de facto foreclosed. Consensus is precisely what suspends the democratic:

> Consensus is thus not another manner of exercising democracy. ... [It] is the negation of the democratic basis for politics: it desires to have well-identifiable groups with specific interests, aspirations, values and "culture". ... Consensualist centrism flourishes with the multiplication of differences and identities. It nourishes itself with the complexification of the elements that need to be accounted for in a community, with the permanent process of autorepresentation, with all the elements and all their differences: the larger the number of groups and identities that need to be taken into account in society, the greater the need for arbitration. The "one" of consensus nourishes itself with the multiple. (Rancière 2000a, 125)

Something similar is at work in the micropolitics of local urban struggles, dispersed resistances, and alternative practices that customarily suture the field of urban social movements today. These are the spheres where urban activism dwells as some form of placebo-politicalness (Marchart 2007, 47). This anti-political impulse works through colonization of the political by the social through sublimation. Such urban struggles identify ruptures, disagreements, contestations, and fractures that inevitably erupt out of the incomplete saturation of the social world by the police order with a political act. For example, the variegated, dispersed, and occasionally effective (on their own terms) forms of urban activism that emerge within concrete socio-spatial interventions, such as, among others, land-use protests, local pollution problems, road proposals, urban development schemes, airport noise or expansions, the felling of trees or forests, the construction of incinerators, industrial plants, mining conflicts, etc. ... elevate the mobilizations of localized communities, particular groups and/or organizations (like NGOs), etc. ... to the status of the political. They become imbued with and are assigned political significance. The space of the political is thereby "reduced to the seeming politicization of these groups or entities. ... Here the political is not truly political because of the restricted nature of the constituency" (Marchart 2007, 47). The identitarian elevation of matters of fact to matters of concern seems, in such a context, to constitute the horizon of the political, of what is possible, of what can be thought and done within the existing configuration. While important and necessary, such interventions operate within the contours of the constituted order. In other words, particular urban conflict is elevated to the status and the dignity of the political. Rather than politicizing, such particularistic social colonization erodes and outflanks the political dimension of egalibertarian universalization. The latter cannot be substituted by a proliferation of identitarian, multiple, and ultimately fragmented communities. Moreover, such expressions of protest that are framed fully within the existing practices and police order (in fact, these protests, as well as their mode of expression, are exactly called into being through the practices and injunctions of the existing order—they are positively invited as expressions of the proper functioning of "democracy") are, in the current post-politicizing arrangement, already fully acknowledged and accounted for. They become either instituted through public-private stakeholder participatory forms of governance, succumbing to the "tyranny of participation" (Cooke and Kothari 2001), or, if they reject the post-democratic frame, are radically marginalized and symbolized as "radicals" or "fundamentalists" and thereby relegated to a domain outside the consensual post-democratic arrangement; they are rendered nonexistent.

In contrast to these impotent passages to the act, the political as conceived in the context of this book is understood as an emergent property discernable in "the moment in which a particular demand is not simply part of the negotiation of interests but aims at something more, and starts to function as the metaphoric condensation of the global restructuring of the entire social space" (Žižek 1999, 208). It is about the recognition of conflict as constitutive of the social condition, and the naming of the spatialities that can become without being grounded in universalizing notions of the social (in the sense of a unfractured community or a sociological definition of "equality," "unity," or "cohesion") or in a singular view of "the people". The political becomes, for Žižek and Rancière, the space of litigation, the space for those who are not-All, who are uncounted and unnamed, not part of the police (symbolic or state) order. A political space is a space of contestation inaugurated by those who have no name or no place.

The elementary gesture of politicization is thus "this identification of the non-part with the Whole, of the part of society with no properly defined place within it (or resisting the allocated place within it) with the Universal" (Žižek 2006b, 70). Such new symbolizations through which what is considered to be noise by the police is turned into speech signals an incipient repoliticization of public civic space in the polis. Reclaiming democracy and the insurgent design of democratizing public spaces (as spaces for the enunciation of agonistic dispute) becomes a foundation and condition of possibility for a reclaimed polis, one that is predicated upon the symbolization of a positively embodied egalibertarian socio-ecological future that is immediately realizable. These symbolizations start from the premise that equality is being wronged by the given urban police order, and are about claiming/producing/carving out a metaphorical and material space by those who are unaccounted for, unnamed, whose fictions are only registered as inarticulate utterances. Insurgency is, therefore, an integral part of the aesthetic register through which the reframing of what is sensible is articulated and can become symbolized. This is a call for a desublimation and a decolonization of the political or, rather, for a reconquest of the political from the social, a reinvention of the political gesture from the plainly depoliticizing effects of post-political and post-democratic policing.

Incipient Urban Politicization

From the Immanence of the Event …

Alain Badiou has explored the significance of these insurrectional events (Badiou 2012). For him, the proliferation of these insurgencies is a sign of a return of the generic ideas of freedom, solidarity, equality, and emancipation (and generically go under the political name of "communism," the historically invariant name for emancipatory struggle—see chapter 8). The historical-geographical experimenting expressed through insurgent activities—that have not (yet) and may never acquire a political name or symbolization (and surely a return to the name of "communism" to designate these movements is unlikely)—nonetheless expresses for Badiou a certain fidelity to the generic communist hypothesis understood as a fidelity to the truth "that a different collective organization is practicable, one that will eliminate the inequality of wealth and even the division of labour. The private appropriation of massive fortunes and their transmission by inheritance will disappear. The existence of a coercive state, separate from civil society, will no longer appear a necessity: a long process of reorganization based on a free association of producers will see it withering away" (Badiou 2008a, 35). A range of observers have systematically commented on and argued for a more in-depth theoretical and practical engagement between the insurrectional movements and the experimental practices that articulate around new forms of egalitarian and solidarity-based management of the commons (see, for example, Badiou 2010; Bosteels 2011; Dean 2012; Douzinas and Žižek 2010; Swyngedouw 2010b; Žižek 2013b).

Badiou regards the recent urban insurgencies as "historical riots" that are marked by procedures of *intensification, contraction,* and *localization* (Badiou 2012, 90–91), markers that also confirm that emancipatory politicization involves the production of its own space. First, intensification refers to the *enthusiasm* marked by an intensification and implosion of time, a radicalization of statements, and an explosion of activities, condensed in an emblematic space that is reorganized to express and mobilized to relay this enthusiasm. All manner of people come together in an intensive explosion, of an intensified process of being that energizes and incites others to share the enthusiasm inaugurated by the event. A politics of encounter, of opening up, and of joining up animates such intensification (Merrifield 2013). Radicalizing statements, actions, and forms of taking side coincide with an intensification of time in place, creating "an active process of correspondence … between the universality of the Idea and the singular detail of the site and the circumstances" (Badiou 2012, 90–91). The tension between the particular and the universal is short-circuiting as the universal starts operating through the transformation of the particular. Such an intense state of collective creation cannot be other that short-lived. Nonetheless, the Idea crystallized in the insurrectional event will last long after the return to the "normality" of everyday life. What is at stake, then, is how to organize the energy and to universalize the Idea inaugurated in the founding event, how to engage in the slow, difficult, and protracted process of inaugurating a new sensibility, a new common sense, of nurturing fidelity, after the initial enthusiasm that marks the historical moment begins to dissipate.

Second, these enormous vital energies are mobilized for a sustained period of time in a contracted manner. All manner of people come together in an intensive explosion of acting, of an intensified process of being-in-common. This intensity operates in and through the collective togetherness of heterogeneous individuals who in their mode of being-in-common, in their multiplicity and process of political subjectivation (that is, in becoming a political actor) and in their encounter, stand for the metaphorical and material condensation of the People (as political category). It is the emergence of a thinking and acting minority that takes the generic position of "the people." In doing so, they "replace an identitarian object, and the separating names bound up in it (like Muslim, Christian, worker, intellectual, young, old, woman, man) with the common name of 'we, the People'" (Badiou 2012, 92).

Finally, a political Idea/Imaginary cannot find ground and grounding without localization. A political moment is always placed, localized, and invariably operative in public space (Badiou 2017). Squares and other (semi-)public spaces, like picket lines, workers' or women's houses, occupied factories, or the Italian *Centri Sociali*, have historically always been the sites, the geographical places, for performing and enacting emancipatory practices; these are the sites of existence, of exhibition, of becoming popular. Without site, a place, or a location, a political idea is impotent. The location produces intensity, unity, and presence, and permits contraction. However, such intense and contracted localized practices can only ever be an event, originary, but ultimately pre-political. It does not (yet) constitute a political sequence.

In sum, the political emerges when the few claim the name of the many, the community as a whole, and are recognized as such. The emergence of political space is always specific, concrete, particular, minoritarian, but stands, in a sort of short-circuiting, as the condensation of the universal. This has to be fully endorsed and the consequences carefully considered. In particular, it pits a democratizing process often against majoritarian, but ultimately passive and objectified, representative democracy. It is worth quoting Alain Badiou here at length:

It is then much more appropriate to speak of popular *dictatorship* than democracy. The word "dictatorship" is widely execrated in our "democratic" environment. ... But just as movement democracy, which is egalitarian and direct, is absolutely opposed to the "democracy" of the executives of Capital's power, which is inegalitarian and representative, so the dictatorship exercised by a popular movement is radically opposed to dictatorships *as forms of separated, oppressive state.* By "popular dictatorship" we mean an authority that is legitimate precisely because its truth derives from the fact that it legitimizes itself. No one is the delegate to anybody else. (Badiou 2012, 59; emphasis in original)

Such movement democracy, minoritarian yet presenting and recognized as the general will of the people, destabilizes liberal notions of instituted democratic forms and forces us to consider "the democratic" as process against democracy as constituted arrangement, or, in other words, to think, with Miguel Abensour, of "democracy against the state" (Abensour 2004).

An egalitarian politics is radically inclusive; everybody is invited in, "it is an inclusionary struggle" (Žižek 2013a, 126). Of course, the question then arises of how to confront those who remain on the outside, who will mobilize whatever dispositive to prevent the universalization of the inclusionary struggle. Against their symbolic and objective violence, it is vital to think about ways to protect and defend the universalizing process without descending into abyssal terror, about how to navigate the prospect of failure in the absence of effective defense as experienced by the Paris Commune or in the violence of political terror that marked so much of past emancipatory transformations. Such violent encounters, of course, always constitute a political act, one that can be legitimized only in political terms. Neither philosophical musings nor substantive social theory can serve to legitimize such encounters.

... to a Sustained Political Sequence

A political truth procedure or a political sequence, for Alain Badiou, unfolds when in the name of equality fidelity to an event is declared; a fidelity that, although always particular, aspires to become public, to universalize. It is a wager on the truth of the egalitarian political sequence (Badiou 2008a). Such a sequence can retroactively be traced through its process of delocalization from or spatialization of the originary site, encapsulated when, for example, the *indignados* claimed, "We are here, but anyway it's global, and we're everywhere." While aspiring to universalize, such spatializing movement can never be totalizing. Indeed, while everyone is invited in, not all will accept the invitation; lines of separation and demarcation will invariably begin to choreograph the politicizing struggle for transformation. The repetition of the repertoires of action, the continuing identification with the originary Idea, and the moving back and forth between insurrectional sites, may begin to tentatively open up new spatialities of transformation while prefiguring experimental relations for new organizational forms. Such a process of spatialization renders concrete, gives content, to the "equality" expressed in the orginary event. In the process, equality becomes substantively embodied and expressed; and alongside with this, a new political name that captures the new imaginary and its associated new commons sense may potentially emerge.

While staging equality in public squares is a vital moment, the process of transformation requires indeed the slow but unstoppable production of new forms of spatialization quilted around materializing the claims of equality, freedom and solidarity. In other words, what is required now and what needs to be thought through is whether or not and how these proto-political localized events can turn into a spatialized political "truth" procedure; it is a process that has to consider carefully the persistent obstacles and often-violent strategies of resistance orchestrated by those who wish to hang on to the existing state of the situation. This procedure raises the question of political subjectivation and organizational configurations, and requires perhaps forging a political name that captures the imaginary of a new egalitarian commons appropriate for twenty-first century's planetary form of urbanization. While during the nineteenth and twentieth century these names were closely associated with "communism" or "socialism" and centered on the key tropes of the party as adequate organizational form, the proletarian as privileged political subject and the state as the arena of struggle and site to occupy, the present situation requires a reimagined socio-ecological configuration and a new set of strategies that nonetheless still revolve around the notions of equality. However, state, party, and proletarian may no longer be the key axes around which an emancipatory sequence becomes articulated. While the remarkable uprisings since 2011 signaled a desire for a different political configuration, there is a long way to go in terms of thinking through and acting upon the modalities that might unleash a transformative democratic political sequence. Considerable intellectual work needs to be done and experimentation is required in terms of thinking through and prefiguring what organizational forms are appropriate and adequate to the task, what the terrain of struggle is, and what or who the agents are of its enactment.

The urgent tasks now for those who maintain fidelity to the political events choreographed in the new insurrectional spaces that demand a new constituent politics (that is, a new mode of organizing everyday environments) center on inventing new modes and practices of collective and sustained political mobilization; organizing the concrete modalities of spatializing and universalizing the Idea provisionally materialized in these intense and contracted localized insurrectional events; and assembling a wide range of new political subjects who are not afraid to stage an egalitarian being-in-common, imagine a different commons, demand the impossible, perform the new, and confront the violence that will inevitably intensify as those who insist on maintaining the present order realize that their days might be numbered. Such post-capitalist politics is not and cannot be based solely on class positions. As Marx asserted long ago, class is a bourgeois concept and practice. The insurgencies are not waged by a class, but by the masses as an assemblage of heterogeneous political subjects. It is when the masses as a political category stage their presence that the elites recoil in horror.

In the aftermath of the insurgencies of the past few years, a veritable explosion of new socio-spatial practices to experiment with has occurred, from housing occupations and movements against dispossession in Spain to new, rapidly proliferating egalibertarian lifestyles and forms of social and ecological organization in Greece, Spain, and many other places, alongside more traditional forms of political organizing. Not all experiments will succeed. Many will fail. In the face of inevitable setbacks—like the current catastrophe in Egypt or Turkey—the fidelity to the democratizing process needs to be maintained and sharpened. Extraordinary experimentation with dispossessing the dispossessor, with reclaiming the commons and organizing access, transformation, and distribution in more egalibertarian ways already marks the return to "ordinary" everyday life post-insurgencies. The incipient ideas expressed in the event materialize in a variety of places and ways, and in the midst of painstaking efforts to build alliances, bridge sites, repeat the insurgencies, establish connectivities, and, in the process, produce organization, symbolize its practices, and generalize its desire. The repetition of heterogeneous situations may well be—as Nick Srnicek argues—what is adequate today to sustain fidelity to the events choreographed by the incipient politicizations of recent insurgencies (Srnicek 2008). Such procedures require painstaking organization, sustained political action, and a committed fidelity to universalizing the egalitarian trajectory for the management of the commons. While staging equality in public squares is a vital moment, the process of transformation requires the slow but unstoppable production of new forms of spatialization quilted around materializing the claims of equality, freedom and solidarity. This is the promise of the return of the political embryonically manifested in insurgent practices.

From Ground Zero to Enacting the Polis

The Real of the political cannot be fully suppressed and—I claim—returns presently, and among others, in the form of the urban insurgencies with which I opened this chapter. If the political is foreclosed and the polis as political community is moribund in the face of the post-politicizing suspension of the democratic, what is to be done? What design for the reclamation of the polis as political space can be imagined and created? How and in what ways can the courage of the urban collective intellect(ual) be mobilized to think through a design of and for dissensual or polemical spaces. I would situate the tentative answers to these questions in three interrelated registers of thought.

The first revolves around transgressing the fantasy that sustains the post-political order. This would include not surrendering to the temptation to engage in the "passage to the act." The hysterical act of resistance ("I have to do something or the city, the world, will go to the dogs") just answers the call of power to do what you want, to live your dream, to be a "responsible" citizen. Acting is actually what is invited, an injunction to obey, to be able to answer the question "What have you done today?" What is required today is the task of envisioning the formation of new ways of commoning and new procedures to inaugurate egalibertarian modes of living.

The second moment of reclaiming the polis revolves around recentering/redesigning the urban as a democratic political field of disagreement. This is about enunciating dissent and rupture, and the ability to literally open up spaces that permit acts that claim and stage an egalitarian place in the order of things. This must include of course the constitution and construction of common spaces as collectivized spaces for experimenting and living differently to counter "the hyper-exploitation of the time that is imposed and that one tries to re-appropriate" (Kakigianno and Rancière 2013, 24). Political space emerges thereby as the collective or common space for the institutionalization of the social (society) and equality as the foundational gesture of political democracy (the presumed, axiomatic, yet contingent foundation).

This requires major thought and practice that cut through the dominant tropes of contemporary governance (including creativity, sustainability, growth, cosmopolitanism, participation) (see Gunder and Hillier 2009). Such metonymic re-registering demands thinking through the city as a space for accommodating equalitarian difference and disorder. This hinges critically on creating egalibertarian public spaces. Most important, the utopian framing that customarily informs urban visioning requires reversal to a temporal sequence centered on imagining concrete spatio-temporal utopias as immediately necessary and realizable. This echoes of course Henri Lefebvre's clarion call for the "Right to the city" understood as the "Right to the production of urbanization," one that urges us to think of the city as a process of collective co-design and coproduction (Harvey 2012).

8 Exploring the Idea of Emancipatory Geographies for the Twenty-First Century

If the forces of wealth and finance have come to dominate supposedly democratic constitutions, ... is it not possible and even necessary today to propose and construct new constitutional figures that can open avenues to again take up the project of the pursuit of collective happiness? [The European and United States protests] pose the need for a new democratic constituent process. (Hardt and Negri 2012)

On the cover of its issue of February 17, 2009, *Newsweek* announced boldly that "We are all socialists now." It referred to the unprecedented nationalization of banks and the multi-billion-dollar rescue packages, demanded by the financial elites and agreed to by both George Bush Jr. and Barack Obama, that the US federal government was pumping into an ailing capitalist economy. And many other governments followed suit during the subsequent years in a desperate attempt to contain the spreading financial crisis. If we take *Newsweek*'s point seriously, the choice we are presented with today is no longer the one Marx once held up, of choosing between barbarism and socialism, but rather between socialism and communism. I am sure the use of the word "communism" in this book may have raised a few eyebrows. The persistent outlawing of the name and its erasure from the collective memory as a possible and reasonable alternative to the present condition over the past two decades or so has been so effective that even its utterance is looked at with suspicion, distrust, and perhaps, a slight sense of curiosity. In an age in which anything and everything can be discussed, the very idea of communism as a positive injunction seems to have been censored and scripted out of both everyday and intellectual vocabularies. It is only tolerated in sensationalized accounts of the "obscure disaster"[1] of twentieth-century socialism in Eastern Europe, or in romanticized Hollywood renditions of the life and work of communist heroes like Che

Guevara. Of course, the work of Karl Marx, Antonio Gramsci, or Louis Althusser is still discussed in arcane academic tracts, but this is much less the case for the political treatises of, say, Vladimir Lenin, Leo Trotsky, Rosa Luxemburg, Mao Zedong, or Ho Chi Minh. The idea of communism has either been stigmatized beyond recovery or relegated to the dustbin of irretrievably failed utopias.

The argument I present in this concluding chapter is not about the lingering remnants of political practices that go under the name of "communism" in various parts of the world (like China, Cuba, North Korea, Nepal, or parts of India). China has decidedly taken the route of authoritarian capitalism and North Korea is nothing more than a tyrannical hereditary dictatorship. Rather, I intend to explore the communist hypothesis as defined by Alain Badiou (see chapter 7) and offer some tentative lines of enquiry related to the possibilities and desirability to revive this hypothesis for the twenty-first century. The key point is not to tease out what the history of communist thought and practice may mean or whether it is still relevant today, but much more important, to analyze how the present conditions look from the perspective of the "Idea of Communism."

Between March 13 and 15, 2009, more than a thousand people convened at Birkbeck College in London to attend a conference on the "Idea of Communism" that gathered about two dozen philosophers from around the world to think about the potential contemporary relevance of the "Idea of Communism." "Move over Jacko, Idea of Communism is [the] hottest ticket in town this weekend," the *Guardian* newspaper announced in its Thursday edition to express its surprise at the success of this sell-out event at about the same time that ticket sales for the planned Michael Jackson concerts started (Campbell 2009). As Alain Badiou noted in the foreword to the conference program:

The communist hypothesis remains the good one, I do not see any other. If we have to abandon this hypothesis, then it is no longer worth doing anything at all in the field of collective action. Without the horizon of communism, without this idea, there is nothing in the historical and political becoming of any interest to a philosopher. Let everyone bother about his own affairs, and let us stop talking about it ... what is imposed on us as a task, even as a philosophical obligation, is to help a new mode of existence of the hypothesis to deploy itself.[2]

The two central metaphors that sustain the idea of communism are, of course, equality and democratic self-governance, held together by a fidelity to the belief that these can be realized geographically through sustained and committed political struggle. The realization of these principles involves the self-organization and self-management of people, and, therefore, will eliminate the coercive state as the principal organizer of political life. Organized and self-confident social and political struggle would be the means by which the former would be realized. The communist hypothesis combines the negativity of "resistance" (to any relation or practice that perverts the presumption of (political) equality) with a belief in the immanent practicability of free and equal forms of socio-spatial organization. However, the political practice of achieving what Badiou calls for seems today more remote than ever, despite the positively verifiable geographical facts that we are moving further away from socially equal and politically democratic forms of "being-in-common" (Nancy 1991) in a geographically highly diversified world. In fact, we are today much closer to the proliferation of new nationalisms and associated exclusionary politics, exemplified by Trump's USA, Brexit UK, and the growing success of xenophobic and radically nationalist or religious movements in many other parts of the world.

The Idea of Communism retains a subversive edge—in spite of the failed experiments that went under that name in the twentieth century—precisely because the name still evokes the sense that a genuinely different world not only is imaginable but also is practically possible. Being-in-common in egalitarian and free ways that permit the self-development of each and all retains a great mobilizing potential. An urgent and demanding intellectual task is required to rethink the socio-spatial practices, the possible forms of political organization, and the transformative imaginaries and material geographies that will give the idea of communism again a positive content. The courage of the intellect needs to be mustered to work through if, how, and in what ways a communism for the twenty-first century can be imagined again. In the remainder of this conclusion, I shall offer some pointers for the formidable intellectual task ahead. I shall proceed in two steps. First, I outline the presumptions on which the communist hypothesis is based. Second, I briefly summarize the political fault lines that striate the current phase of post-political democracy and neoliberal capitalism. On the way, I shall chart tentatively what is left to think.

The Idea of Communism

Today, communism is just an idea, a hypothesis, and a scandalous and illegitimate one in the present sequence of things. The question is whether new and different significations can be inscribed in its name, rather than endlessly repeating the truth of the ethico-moral bankruptcy of the former Communist bloc. Neither a rehearsal of the standard critiques of once really existing communism nor a simple invocation of its utopian possibilities will do; what is required is nothing less than a radical invention of the new, on the basis of a sustained critical engagement with what was and is already embryonically there. It is useful, in this context, to recall Marx's definition of communism from *The German Ideology*: "Communism is for us not a *state of affairs* which is to be established, an *ideal* to which reality will have to adjust itself. We call communism the *real* movement which abolishes the present state of things" (Marx and Engels 1987, 56–57; emphasis added). Tentatively charting the contours of this real movement for the twenty-first century is the challenge for the contemporary collective communist intellect.

Communism is intimately connected to the democratic. I am not referring here to democracy as a set of political institutions (parliaments, governments, and the like) and its associated political procedures (like elections at regular intervals in which a recognized set of individuals can participate if they wish to do so), but rather to the founding gesture of democracy. The democratic political, as explored in chapter 3, expresses the contingent presumption of equality of each and every one qua speaking—and hence political—beings. The contingent presumption of equality that marks "the democratic invention" stands in strict opposition to any given, and sociologically verifiable, order, including any given and instituted "democratic" order. The democratic political, therefore, exposes the unegalitarian processes that rupture any given socio-spatial order. In other words, equality is the very premise upon which a democratic politics is constituted. Justice, from this perspective, disappears from the terrain of the moral and enters the space of the political under the name of equality. For Etienne Balibar, the unconditional premise for justice and emancipation resides in the fusion of equality and liberty (what he names as "égaliberté") (Balibar 1993). However, neither freedom nor equality are offered, granted, or distributed, they can only be conquered. The democratic political, therefore, is not about expressing demands to the elites to rectify injustices, inequalities, or unfreedoms, but about the enunciation of the right to *égaliberté*. The idea of communism is thus premised on the unconditionality of equality in a given institutional and social arrangement that has always already "wronged" the very condition of equality (Rancière 2001). Put simply, a communist political sequence arises when, in the name of equality, those who are not equally included in the existing socio-political order, demand their "right to equality," a demand that calls the political into being, renders visible what is invisible, and exposes the "wrongs" in the present order.

Moreover, the presumption of equality is predicated upon asserting difference, differentiation, agonism, and dispute, while refusing to inscribe one particular antagonism as the One that prevents the realization of the presumption of equality. Finally, the presumption of equality assures that the place of power is kept structurally vacant (Lefort 1986) or, in other words, anyone can claim the place of power. There is neither a transcendental figure (like "the King") nor a universal political subject (like "the Proletariat" or "the Party") that can and should suture the place of power. A democratizing political sequence, then, is of course one that demands equality in the face of clear and present exclusions that are part of any (democratic or otherwise) instituted order. A communist practice is one that struggles for the positive realization of equality (as a historically-geographically contingent and, therefore, always contestable inscription, one that demands enduring verification and reimagination) in the face of inegalitarian practices, and strives for the universalization of this egalitarian injunction from the basis of always historically and geographically situated and locally specific inequalities. The communist hypothesis offers a testing ground for the "truth" of processes of changes that operate under the signifiers of "equality" or "emancipation."

Consider, for example, the emancipatory-egalitarian struggles of the proletarian subject in the nineteenth century, demanding political equality in a republican configuration that disavowed the persistent perversion of the egalitarian principle, which was nevertheless enshrined in the republican constitution. Or the struggle of women in the twentieth century for political recognition, the extraordinary fights of African Americans and the South African ANC for egalitarian democratic political emancipation. Today's struggles of immigrants (undocumented or otherwise), demanding political equality, equally express the egalitarian democratic desires that underpin the communist hypothesis. It is these demands that the instituted democratic order with its allocation of places and functions systematically perverts in all manner of ways (Swyngedouw 2009b). These struggles are invariably located in concrete places but aspire to universalization and spatialization; they are predicated on political subjectivation, the becoming of a political subject through grounded emancipatory struggles. The presumption of equality that operates under the name of "democracy" is of course an integral part of the idea of communism. Rethinking and reclaiming the political notion of democracy as equality is a central and vital task.

However, the demand for political equality is a necessary, but not sufficient, condition for the realization of the communist hypothesis. Political equality prefigures and permits the agonistic expression of differential claims, particularly with respect to the forms of social organization and the distribution of collective wealth (or surplus). Political equality assumes the capacity of each and every one to govern, and affirms the capability of self-organization and collective decision-making. This opens up a second terrain—after equality—that sustains the communist hypothesis, namely the fateful belief in the capacity of everyone (and not just of the state, its technocratic managers, or propertied elites) to govern and to decide the principles of appropriation, mobilization, and distribution of wealth and revenue. The communist hypothesis, therefore, prefigures the end of the coercive state as we know it and its replacement by forms of self-organization and self-management within a constellation of multi-scalar governance. Thinking through the relations between emancipatory struggles and the transformation of forms of governing the commons is indeed an urgent task. In particular, the articulation between different interlocking scales of regulation, self-management, and organization on the one hand and their relation to changing state forms on the other remains a thorny issue.

The historical-geographical terrain for the realization of the communist hypothesis is of course the commons. "Commonism" is the substantive ground on which the hypothesis realizes its promises. The very name of "communism" not only invokes an egalitarian "being-in-common" of all qua multiple and multitude, but also includes the commons that is the earth, the world, and therefore life itself. This latter sense of the commons refers fundamentally to the collectively transformed socio-ecological relations, such as water, air, and CO_2, but also knowledge, information, affective labor, biodiversity regimes, resources, urban space, and the like. The concept of the commons includes both the nonhuman that has become an integral part of the assemblages that we inhabit and the social bond that inscribes our bodies in a common existence. The commons of socio-ecological arrangements and their conditions of rights of use, transformation (metabolization), access and distribution, the modalities of their spatial organization, and the configuration, access (education), ownership, and distribution of collective knowledge/information is now the key domain around which the communist hypothesis has to be thought and developed. In particular, it raises the question of property and property relations with respect to common resources like those I have exemplified. Cloaking the political argument around the commons as one between public versus private property, I would argue, misses the point if the public sphere is defined as or restricted to the domain of the state. As I shall argue further below, the state has become (and arguably has always been) another instance of the private, distinct from private capital or individuals, but nevertheless pursuing what Immanuel Kant would call private reason with respect to the commons and to the ownership of the biopolitical conditions of life like the environment, resources, genetic and informational code systems, knowledge, and so on. The communist hypothesis is structured around the commons as the shared ownership of each and every one under common stewardship. Communism, therefore, is a struggle against both the privatism of the state and that of capital—ultimately sanctioned by property relations that fragment, privatize, and monopolize the commons—and for the production of collective institutions for the democratic management of the commons, thereby turning the commons into a new use value that cannot be turned into exchange value. The communist idea, therefore, is also about the transformation of the commons from private to collective, the abolition of private property (of the commons), and reaffirmation of the capacity of all qua collective—the communist intelligence to govern the city, the commons-in-common.

The struggle over and for the commons also highlights the now irreconcilable difference between current forms of social-democratic socialism (what Paul Virno defines as the socialism for capital, to follow) and communism. As Toni Negri put it: "The need to distinguish between 'socialism' and 'communism' has again become obvious: but this time not because of the blurred boundaries between them, but because they are so opposed. Socialism is nothing other than one of the forms taken by capitalist management of the economy and of power, whereas communism is an absolutely radical political economic democracy and an aspiration to freedom" (Negri 1990, 167).

The communist hypothesis contains a historical invariant—it stands for the eternal return of the emancipatory struggles sustained by the recurrent appearance, albeit in different historical and geographical forms, of the desire/struggle for emancipation, freedom, and equality. Emancipatory struggle is a geographically constituted historical invariant whose traces can be identified throughout the ages, from the struggle of the noncitizens to become part of the *Demos* in ancient Greek city-states, to slaves fighting for freedom or serfs for liberty, to indigenous communities working for their right to sustain their livelihoods. Realizing the communist hypothesis entails a voluntarist (subjective) moment to revive this communist invariant, that is the will of the individual to join up with others to realize politically the idea of communism (Hallward 2009). The communist idea is nothing without the will to do something new, without the will to become a political subject. It insists on the continuous transformation of this singularity of the egalitarian will and movement to the universalizing multiplicity of being-in-common as part of a commons.

Needless to say, the communist hypothesis is confronted with the existing state of affairs, the conditions of contemporary neoliberalizing capitalist forms with their often extreme and multiple socio-spatial and socio-ecological inequalities, uneven political and economic development, extraordinary exclusions, and autocratic global transformations, animated by the joint dynamics of a multi-scaled state and quasi-state forms (like the European Union) on the one hand and the competitive struggle of individual capitalist actors, organized within networks with varying spatial geometries, for surplus-value production and realization on the other. Working through the communist hypothesis requires engaging seriously with this state of the situation and, in particular, with the fate of instituted post-democracy, the types and contours of the multiple emancipatory struggles that mark the landscape of uneven development, and the changing conditions of the commons in the relentless transformations of capitalism. In other words, a critique of political economy, reframed as political ecology, is still urgently needed, a critique that also accounts both for the resurgence of emancipatory and egalitarian desires (the continuous

reemergence of the communist principles) and for the intensifying struggles over the privatization and dispossession of the commons.

The New Spirit of Capitalism: Capitalism's Revolutions

Cultures of Excess

Coming to terms with the "obscure" disaster of twentieth-century communism not only necessitates a communist critique of the uneven and truncated failures of the socialist projects in the Soviet Union or China, among
others, but also of the failure (in emancipatory terms) of state control/man-
agement and of the repressive (in ethical, cultural, and moral terms) state-
capitalism in the West (including many of the post-colonial states) during
the golden age of post-war capitalism. Only an in-depth understanding of
the failures of the socialist management of capital may permit grappling
with how the "New Spirit of Capitalism," which reinvigorated the capitalist class project after the 1970s, permitted the capture of imaginaries
and fantasies that had galvanized so much of earlier communist thought
and political practice (emancipation, freedom, equality, and cosmopolitan
internationalism). The 1968 events, both in the West and the East (like
in Prague or Beijing), were of course as much a revolt against the stale,
repressive, and reactionary moral order of capitalism as against economic
exploitation choreographed by capitalist relations or the excesses of state
domination under actually existing socialism. Indeed, the bureaucratized
"Fordist-welfarist" forms of capitalism—namely, mass production of consumer goods combined with a distributional welfare state—were paralleled by a repressive moral order shared both by the social-democratic and
socialist variants of state management. The revolts of the late 1960s and
early 1970s targeted, among other demands, the suffocating moral restraint
in sexual, gender, money, relationships, and other affective registers that
had been an integral part of the capitalist socio-cultural coding and order
until then. The revolutionaries of the libertarian and romantic currents of
the late 1960s demanded (successfully) all manner of affective liberations.
Conservative constraint was replaced by the imperative to enjoy. The Left's
critique of everyday life in the stale suburbanized living of the diffuse spectacle of western capitalism, as well as in the Stalinist bunker spaces of living in the concentrated spectacle (see Debord 1967), became exquisitely

incorporated—and depoliticized—in what Boltanski and Chiapello call *The New Spirit of Capitalism* (Boltanski and Chiapello 2007; Sennett 2007). Both economic elites and radical cultural critiques and practices reveled in embracing new forms of excess. Liberation was experienced as the search of surplus value as well as surplus enjoyment. The injunction to enjoy became the cultural expression of the latest round of capitalist transformations (Žižek 1999, 2006c); an injunction that also liberated the traditional elites of the traditional bourgeois injunction to care.

Political equality weakened as a central concern. The new capitalist cultural-political economy of excess fused seamlessly with demands for equality that became defined as the equality of difference (and justice framed as the right to be different or, rather, as the right to enjoy one's differential specificity). Democracy, in turn, became defined and pursued as the freedom to exercise individual choice and preference rather than the unconditional given of each and every one as equal qua speaking beings. Consuming became the highest freedom and expression of emancipation. Indeed, consuming identity became the hard kernel around which enjoyment, freedom, and difference circulated. The latter became secured in a market exchange that treats everyone as equal, while rendering everything commensurable under the money sign. Market equality replaced political equality and became metonymically imagined as a superior form of "democracy." Hans Tietmeyer, in 1998 governor of the Deutsche Bundesbank, put it very succinctly when he applauded governments for taking their cue from "the permanent plebiscite of global markets" rather than from the "plebiscite of the ballot box," thereby suspending the political process to replace it with the equality of "the markets" (quoted in Žižek 2017, 17).

Equality as difference and freedom as the enjoyment of excess became the cultural expression of new forms of fragmented, networked, diffuse, and multi-centered capitalisms (Hardt and Negri 2001). Indeed, cosmopolitan global capital can easily and enthusiastically enroll all manner of different cultural values and identitarian inscriptions within its circuits of capital accumulation. A presumably nonauthoritarian and inclusive capitalism, which skillfully displaced—both geographically and organizationally—the more overt mechanisms of repression, exploitation, exclusion, and submission to the global South, hegemonized the signifiers that once belonged to communism. Some of the Left's 1960s critique became incorporated in and subsumed under a revolutionary capitalism. Indeed, not the communist idea and practice, but capitalist "perestroika" became the most revolutionary and exhilarating game in town. Mao's call for permanent revolution that was smothered in China's transition to capitalism while being the condition of the latter's possibility (Russo 2006) had been taken up by the newly emerging (and older transforming) western elites in a class-project to revamp capitalism and reaffirm capitalist class power. This "cultural" transformation of capitalism signals nothing less than a passive revolution of the kind that Gramsci identified in the early 1930s with the emergence of "Fordism," a form of neutralization of originally counter-hegemonic demands by new organizational and managerial forms of capitalism (Mouffe 2000). The fall of the Berlin Wall in 1989 indicated the final victory of the liberal-capitalist model that—now freed from the shackles of a once formidable alternative model—could steam ahead unopposed and reassert more forcefully its inegalitarian project of earthly transformation.

Capturing the State: Socialism for the Elites

Indeed, the onslaught of neoliberalism and the deepening process of neo-liberalization, which is the signifier that stands for the successful class struggle of the bourgeoisie to regain the upper hand over the dispossessed of the world, have all but wiped out working-class politics. As Harvey contends, "[neoliberalism] is a class project, masked by a lot of neo-liberal rhetoric about individual freedom, liberty, personal responsibility, privatization and the free market. These were means, however, towards the restoration and consolidation of class power, and in that the neo-liberal project has been fairly successful" (Harvey 2009, see also Harvey 2005). This class victory is now plain to see in the present conjuncture marked by intensifying sequences of financial-economic crises. In the 1970s, the then-dominant Marxist political theory that conceptualized the state as the executive branch of the capitalist class required sophisticated reformulation (most notably in the work of Nicos Poulantzas and Bob Jessop) to tease out both the increasingly obscured class character of the Keynesian welfare state as well as the possibilities the state offered for socialist transition. Today, there is no longer anything obscure about the mission of state intervention. Bankers on the verge of bankruptcy successfully call on the national state for immediate, urgent, and desperate measures. Unprecedented public deficits are produced for future generations to carry. All manner of palliatives that fit the market rationale are devised to ensure that things can go on as before. This, of course, short-circuits the possibilities for alternative state policies, while those individuals losing their homes, jobs, or whatever flimsy security they had largely remain in the cold, easy targets for a right-wing populism that promises a social-democratic, but deeply nationalist and often xenophobic, return to the relative securities of yesteryear. All of this has now become a process that is fully transparent and legible, barely wrapped in a general discourse of protecting the "common" interest.

This is also discernable, for example, in the extraordinary process of accumulation by dispossession orchestrated through the state in the name of salvaging a capital-financial system built on spiraling fictitious capital formation and circulation. Without much organized protest, financial capital basically took command of the state's capacity and forged it in its own interest, a political coup d'état that indeed returned the state to be the executive managers of the collective interests of the economic elites. This is a return of socialism and of the socialist state, but a socialism that ensures the interests of capital (see Virno 2004), and certainly not the interests of those at the bottom of the pile. The creation of a permanent state of exception indeed has now become an integral part of the normal function of the state (Agamben 2005). If the emergent post-crisis order is structured around socialism for capital, then the future is not about the political choice between market and state, but rather between socialism and communism.

I contend that this "privatization" of the state renders any political project articulated around the national state problematic (albeit still worth pursuing). As I have argued elsewhere, neoliberalization processes are accompanied by multi-scalar reorganizations of the mode of governing, which reorders socio-political power geometries as well as the institutional modalities of governing, toward a form of governmentality that banishes the democratic supplement, resulting in more autocratic (quasi-) state forms (Swyngedouw 2000; see also chapter 1). This suspension of the properly democratic and the consolidation of post-democratic forms of institutional arrangements are supported by post-political tactics of depoliticization. Contemporary capitalism is indeed increasingly authoritarian, as Peter Sloterdijk maintains, with a choice between China's "party dictatorial" mode, the Soviet Union's "state dictatorial" mode, the United States' "sentiment dictatorial" mode, and finally the "media dictatorial" mode of which Berlusconi's Italy became the iconic expression (Sloterdijk 2005).

Financializing Everything and Dispossessing Natures/Bodies

The procedures of neoliberalization do not only constitute simple class tactics of mobilizing the state for the class project, but also signal a profound transfiguration of intra-capitalist relations. There is a clear shift from industrial/commodity-producing capital to financial capital, a process usually referred to as "financialization." While there is considerable dispute over what constitutes financialization, I take it to be a condition whereby the accumulation process is increasingly sustained by the circulation of capital through all manner of financial transactions, rather than by commodity production (see Krippner 2005). The accompanying changing power relations are clearly expressed in the current climate of economic crisis and, in particular, in the variegated ability to mobilize state power by different fractions of capital. It is vital to recognize that financial capital is, of course, not separate from other circuits of capital accumulation. The key is to grapple with their mutually constituted condition of possibility, their interrelationships and tensions. In light of this, the new forms of mobilization of nature and its incorporation within the circuits of capital through new forms of articulation between financial and "real" capital are urgently required.

I would argue that one of the central transformations in the political ecology of contemporary capitalism resides in the changing dynamics and new characteristics associated with property and property relations. In *The Economic and Philosophical Manuscripts*, Marx already alluded to the tensions and struggle between two forms of capital, that is, between immobile versus mobile capital (Marx 1967). The former is land- and resource-based, and surplus is accumulated primarily through various forms of rent extraction and based on property relations rather than accumulation based on the production of surplus value within the valorization process. While the latter is productive of value through the mobilization of labor, the former is extractive in terms of transferring labor values into rents or interest or both. In the past few decades, financialization has indeed accelerated the reversal of inter-capitalist relations back to a greater role of surplus generation through rent extractions of a variety of kinds. Financialization as a particular form of circulating capital, premised upon transforming geographically specific, relatively fixed, and particular conditions into abstract circulating fictitious and interest-yielding capital, has become a key form of what David Harvey calls "accumulation by dispossession." Capitalism's spectacular resurgence, strongly related with this extraordinary reassertion of rent/

interest-yielding "stuff," centered on land-based speculation, the privatization of environmental commons like water, gene pools, CO_2, minerals, and
the like, intellectual property regimes, bio-genetic ownership, affective and
cognitive labor (like software code), and so on (Andreucci et al. 2017).

This rent/interest extraction-form of financial capitalism also increasingly relies on the mobilization and appropriation of collective or common immaterial labor or, as Michael Hardt calls it, biopolitical capital
(Hardt 2010):

> In the final decades of the twentieth century, industrial labor lost its hegemony
> and in its stead emerged "immaterial labor," that is, labor that creates immaterial
> products, such as knowledge, information, communication, a relationship, or an
> emotional response. ... As an initial approach, one can conceive immaterial labor
> in two principle forms. The first form refers to labor that is primarily intellectual
> or linguistic, such as problem solving, symbolic and analytical tasks, and linguis-
> tic expressions. This kind of immaterial labor produces ideas, symbols, codes, texts,
> linguistic figures, images, and other such products. We call the other principle form
> of immaterial labor "affective labor" ... [it] refers equally to body and mind. In fact,
> affects, such as joy and sadness, reveal the present state of life in the entire organ-
> ism, expressing a certain state of the body along with a certain mode of thinking. ...
> Affective labor, then, is labor that produces or manipulates affects. ... A worker with
> a good attitude and social skills is another way of saying a worker adept at affective
> labor." (Hardt and Negri 2004, 108)

These are the sorts of labor that produce reproducible goods like information, codes, affects, designs, images; they are reproducible in a biopolitical sense—they can be produced (albeit in new and always changing
ways) through the biopolitical production of life itself, hence Hardt and
Negri's insistence on defining these forms of immaterial labor as "biopolitical" capital. New forms of property arise from that, usually referred
to as "intellectual property" rights. While formal property of affective
"goods" is difficult to establish, all manner of dispossession tactics, outside of the economic sphere but articulated through legal, state, and other
institutional arrangements, are operative, opening up a vast terrain of
tension, continuous contestation, and occasional subversion (code shar-
ing, pirating, hacking, social economy initiatives, and the like). Software
code, information networks, images, knowledge, smiles, good teaching,
the choice of music on Spotify or even Google searches are difficult to
privatize directly and hence need extraordinary mechanisms to enroll
them with the circuits of market exchange. What is vital to distinguish

here is that the product is "immaterial," not the labor. Financialization and direct dispossession (through violent regimes of establishing private property rights) have become the key tropes through which the common intellect of affective labor becomes incorporated and reproduced within the circulation of capital.

The spiraling forms of immaterial labor upon which much of contemporary capitalism (particularly in the Global North) rests opens up new forms of class conflict that do not focus on the ownership of the means of production, but directly on the ownership of the products of affective labor, mediated through the monetary nexus and the right to the rents produced through collective affective/immaterial labor. All manner of political conflict, both symbolic and material, revolves around the control and expression of these forms of affective labor and the surplus they produce. In other words, social struggle unfolds increasingly around the collective/commons versus the private character of affective capital. An interesting example is the current mobilization of students and academics around the privatization of knowledge, centered on the demand that publicly funded research should not be published in outlets that are privatized and through which public access is restricted by a paywall.

Both the financialization of space/nature (land and other socio-ecological "resources") as well as the financialization of reproducible biopolitical affective goods are the dominant forms of the new culture of capital, conditions that require urgent attention as they signal a nodal point in thinking through the idea of communism for the twenty-first century. As Marx and Engels already observed in *The Communist Manifesto*, expropriation or dispossession is of course not the objective of a communist regime. It is capitalism's expansion that is predicated upon expropriation and dispossessing: in 2016, sixty-two individuals in the world owned as much wealth as the poorest 50 percent of the global population! This number fell to eight individuals in 2017. Reclaiming ownership of the privatized commons under collective management is, of course, an integral part of the communist demands.

The flipside of biopolitical production is of course immunological biopolitical management, that is, the management and regulation of the security, immunity, and welfare of the people, the management of fear or, more precisely, the management of the fear of fear (see Graham 2009). It is in this form that the properly ideological regime of the new spirit of capitalism is directly evident. Whether the fear of globalization and loss of competitiveness, fear for an ecological Armageddon and climate collapse, fear of political Islam or the flood of (il)legal or undocumented immigrants, the cultivation of the fear of fear is the central trope through which the integrity of the state is maintained while, at the same time, the apocalyptic imaginary that accompanies these discourses is one without the promise of redemption, one that can neither be fulfilled nor overcome, only postponed (see chapter 5). The politics of fear are central to the post-politicization or depoliticization that has characterized the past few years (Badiou 2008c). The communist hypothesis radically breaks with fear—communism is about the faithful belief and relentless call for and struggle over the staging of equality and freedom, sustained by the conviction that these are immediately realizable and immanently practical.

Revolutionized Uneven Geographies of Capitalism

The support structures that permit the proliferation of affective labor are deeply material, albeit geographically unevenly organized. In the contemporary new spatial divisions of labor, material production is increasingly (although by no means entirely) carried out in China, India, and other newly emerging capitalist spaces, while resource grabbing and exploitation choreographs much of the political-ecological transformations in Africa and parts of Latin America (Andreucci et al. 2017). Forms of direct labor exploitation, based on ultra-Taylorist modes of labor organization under severe and multiple hyper-exploitative conditions, have become the material engines of world production. While the proletariat as a political force may be disappearing in the Global North, the proletariat, both sociologically and politically, is being constituted on an extraordinary scale in the new spaces of capitalism, and with it, growing labor unrest and conflicts as the following example of China illustrates (see also the China Labour Bulletin at http://www.clb.org.hk/,accessedJuly3,2016):

By 2006 the Chinese working class numbered 764 million, with 283 million living in the cities. Bitter labour disputes, anger at rising food prices and pollution have taken a growing number of people onto the streets. In 1994 an estimated 740,000 joined street demonstrations. By 2004 the number rose to 3.7 million. Strikes rose from 1,909 in 1994 to 22,600 in 2003 and the number of strikers has grown from 77,704 to 800,000. … 30,000 workers in over a dozen factories in the Dalian Development Zone struck in 2005. (https://www.solidarity.net.au/, accessed April 7, 2009)

In addition, much of the affective labor (care, catering, security, and the like) and its material support structures in the Global North are organized through international migrant labor, often sustained by undocumented immigration, refugees, or other precarious ways of life. The (il)legal migrant has become the ideal deregulated (undocumented) body as mere laborer. An extraordinary form of inter- and intra-national exclusion and inequality is structured through these new affective forms of labor organization and appropriation. Both these emergent forms of political subjectivity require urgent theoretical-political attention.

Globally speaking, therefore, the demands for equality and the staging of freedom will in all likelihood articulate based on the proletarian political subject in the spaces that are configured as the "assembly lines of world," while struggles around common property, dispossession, citizenship rights, control of affective, biopolitical equality, and the like are emerging and proliferating in the Global North as well as among many of the structurally dispossessed who live in the slums of the world's megacities. In sum, the new faces of capitalism form increasingly around processes of changing property relations, articulating both cognitive/affective and material commons, and primarily organized and choreographed by financialization and its process of rent/interest extraction.

However, the recent and ongoing financial crisis shows the ineffectiveness, if not inability, of capital to govern or manage the commons, signaling a growing contradiction between the conditions of private property on the one hand and the organization and mobilization of the commons on the other. Communism is radically concerned with foregrounding the commons and the abolition of the exclusive private property of the commons upon which contemporary capitalism rests or, in other words, communism is about the democratic management of the commons.

This discussion of the commons brings us directly to one of the vital, but ultimately fundamentally disavowed conditions of our time, which is the ecological quagmire. The ecological question relates directly to the commons, our common life world. The attempts to mainstream the ecological problem are marked by three interrelated processes. First, the apparent greater ecological sensitivity of capital and of the elites; second, the nurturing of a particular discourse of the environment, and third, the continuing financialization/privatization of the environmental commons, whether in the form of CO_2, bio-pools, water and other resources, human and nonhuman genetic codes, and the like, including of course the gigantic privatization of the greatest of all common ecologies, the urban process. The "environmental wedge," as Noel Castree suggests, reorients the political mindset (Castree 2009). For the first time in history, human/nonhuman interactions may produce socio-ecological conditions that undermine the continuation of human and other life forms. Although we may not know all the ecological consequences of human's socio-ecological labor, the possibility of interventions to spiral outward into too risky a terrain has now become reality. Yet, capitalism cannot and will not stand for an unconditional demand for a transformation to a different egalitarian socio-ecological order, despite the call to arms from a variety of elites, ranging from Prince Charles's apocalyptic warnings about the dangers of climate change to Al Gore's biblical *An Inconvenient Truth* (see chapters 4 and 5). The fantasy that immediate and urgent action would indeed be taken as a result of these dystopian forebodings was of course terminally shattered by the unconditional demand during the 2008–2010 financial crisis to take immediate and urgent action to "save the banks" (other demands, most notably the alleged pending environmental catastrophe, can wait). Democracy was instantaneously suspended as executive measures taken by both national governments and transnational organizations like the European Union trumped proper democratic control, and the environment disappeared from the top of the agenda as the Real of capital imposed its own urgency. The restoration of capital was predicated upon restoring the fantasy of confidence in "the system" and making money circulate again as capital. The inability or incapacity to manage the commons of socio-ecological assemblages, probably not even in the elites' own interest, is an extraordinary situation. The point, here, is not to fall into the urge to save nature—which does not exist anyway as a stable marker or reference—or to retrofit socio-ecological conditions to an assumedly more benign earlier historical condition (which is of course an inherently reactionary demand), but rather to call for an egalitarian and democratic production of socio-ecological commons.

Moving Bodies/Fixed Bodies

What the new spirit of capitalism points at is the general privatization of the commons: the commons of the intellect/affect, the commons of external nature, and the commons of internal biogenetic nature. The unprecedented enclosure of the registers of the commons through privatization points to a final and, arguably, crucial conflict and contradiction, namely the dialectic of inclusion and exclusion: the separation between those who are part and those who are not. The figure that literally inscribes the markers of the proliferation of walls, demarcations, separations, and the multiple insides/outsides of the current geopolitical order is the moving body and, in particular, the body of the undocumented immigrant, the refugee, the idealized neoliberal subject, the one without political inscription, without papers (and therefore without rights):

Nowadays, when the welfare state is gone, this separation between citizens and non-citizens still remains, but with an additional paradox that non-citizens represent the avant-garde within the neo-liberal project, because they are indeed positioned within the labor force market without any kind of social rights or state protection. Thus, if we examine this problem in such a way, the *sanspapiers* and the *erased* are the avant-garde form of sociality which would prevail if the neo-liberal concept is to be fully realized, if it would not be important anymore if someone is a citizen or not, if everybody would be defined only according to their position in the labor market and the labor process. (Pupovac and Karamani 2006, 48; emphasis in original)

While running the risk of unacceptable overgeneralization, the refugees of Darfur, the Albanese diaspora in Greece and Italy, mass illegal migration from Africa to Europe, the Afghani and Syrian exodus, the Latino migration to North America are all marked by distinct geographies of exclusion and encampment. This undefined "rabble," those who are nonexistent in proper political terms, *homines sacri*, those who do not have a voice to speak, do not have the right to be, yet are everywhere, often itinerant, are the signifier par excellence of the travesty of actually existing "democracy." The flip side of these itinerant bodies are those who cannot move, imprisoned behind walls: material (as in Gaza and the West Bank), political (like those trapped in the zones of continuous tribal warfare), or symbolic (those with no papers or the wrong ones), or concentrated in slums, favelas, asylum centers, or labor camps (like in China). This large and growing army of excluded stand in for the "scandal of democracy," the fact that indeed not everyone is equal and the biopolitical state is here the central demarcating agent of immunitary inclusion and/or violent exclusion as flip sides of the same process. The new walls and lines that demarcate insiders from outsiders into separate—yet intertwined—worlds are very different from the walls of the twentieth century (like the Iron Curtain, symbolized most dramatically by the Berlin Wall). While the latter demarcated a separation between two different and mutually exclusive world views, the new walls of the twenty-first century are erected to keep some out while providing an illusionary immunological prophylactic for the insiders.

A radical democratic demand, therefore, is the one around which undocumented immigrants as political subjects rally: "We are here, therefore we are from here." Of course, this egalitarian demand does not only pertain to the place of utterance, but to all other places to which these multi-scaled bodies "belong" as well (see Swyngedouw and Swyngedouw 2009). Under conditions of abject exclusion, violence can become the only conduit for voicing radical discontent. Indeed, we cannot ignore the rise of subjective violence over the past few years: the burning French *banlieues* (see Dikeç 2007), the rioting students and other youths in Greece's main cities in December 2008, the food riots that spread like wildfire in mid-2008 (in the midst of a massive hike in both food and oil prices), the sequence of urban rebellions in places as different as Italy, Denmark, Moldavia, South Korea, and Haiti, and the like, or the string of ritualistic anti-globalization or climate protests and their perennial promise of violence. The resurgence of such forms of subjective violence, in other words, when participants engage voluntarily in acts recognized as violent and that Mustafa Dikeç dubs "Urban Rage" (Dikeç 2017), seems to be a permanent feature of the new geographies of a post-political world. Subjective violence is of course always measured with respect to a state of apparent nonviolence, a benign condition of absence of violent conflict. This absolutist measuring rod disavows the multiple expression of objective violence, that is the desubjectivized normal condition of institutionalized everyday violence, often of the most brutal and repressive kind (see Žižek 2008c). Consider for example the death toll in Iraq, the genocidal march of HIV in Sub-Saharan countries and parts of Asia in the absence of accessible retroviral drugs, the death by drowning of an unknown and usually unnamed number of refugees that try to reach the shores of Europe or the United States. Or the fact that 1.5 billion people worldwide do not have access to water, a situation that is the world's number-one cause of premature mortality, of people dying before their sell-by date has passed. In the less dramatic circumstances of the Global North, one can think of the violence inflicted by the repossession of homes, rising unemployment, disappearing savings as pension funds fail or banks go bust (García Lamarca and Kaika 2016). These forms of objective violence, normal everyday conditions in the existing state of the situation that are not measured against a condition of nonviolence, are the other side of the coin of the regular outbursts of subjective violence manifested in all manner of rebellions and riots. Universally condemned by the political elites, these are desperate signs and screams for recognition that express profound dissatisfaction with the existing configuration, while testifying to the political impotence of such gestures and signaling the need for a more political, that is politicized, organization of these anarchic expressions for the desire for a new commons.

What Is Left to Think?

We are living in times that are haunted and obsessive in equal measure. On the one hand, our time is haunted by the specter of a once existing but failed communism. The idea of communism is indeed tainted by the failure of its twentieth-century manifestation, a condition that, toward the end of that century, left the Left in a state of utter paralysis, politically and intellectually. This is not to say that the "obscure disaster" of twentieth-century communism does not require urgent and critical attention. On the contrary, this is one of the tasks ahead, one that has to be undertaken in light of the communist hypothesis. But in equal measure are we living in obsessive times, obsessive commitments to "do something," "to act" in the names of humanity, cosmopolitanism, anti-globalization or alter-globalization, the environment, sustainability, resilience, climate change, social justice, or other empty signifiers that have become the stand-in both to cover up for the absence of emancipatory egalibertarian political fantasies and to disavow or repress the socio-ecological antagonisms that structure the reality of the situation. The failure of such obsessive activism is now clearly visible. Humanitarianism is hailed to legitimize military intervention and imperial war, the environment becomes a new terrain of capitalist accumulation and serves ideologically as a "new opium of the masses," cosmopolitanism is cherished as the cultural condition of a globalized capitalism, and anti-globalization manifestations or insurgent rebellions have become the predictable, albeit spectacular, short-lived Bakhtinian, but politically impotent, carnivals whose geographical staging is carefully choreographed by the state. Communist thought has disappeared, to be replaced by relentless, yet politically powerless, resistance (rather than transformation), social critique, and obsessive but impotent acting out.

The relationship between our critical theories and the political as egalitarian-emancipatory process has to be thought again. It is undoubtedly the case that the three key markers of twentieth-century communist politics—state, party, and proletariat—require radical reworking. I would insist that neither the twentieth-century form of the state nor the hierarchical party is of use any longer as privileged terrain and strategy to test the truth of the communist hypothesis. This should not be read as an invitation to ditch forms of institutional and political organization. On the contrary, it calls for a new beginning in terms of thinking through what institutional forms are required at what scale and what forms of political organization are adequate to achieve this. The notion of the proletariat as a political subject equally needs radical overhaul in light of a new critique, not of political economy, but of political ecology. In a context of mass dispossession and privatization of the commons, the political fault lines become centered on this axis around which all manner of new political subjectivities arise. The name of the proletarian stands of course for the political subject who, through egalibertarian struggle, aims to take control again of life and its conditions of possibility. The "proletarian" is therefore not any longer defined by the social position he or she takes within the division of labor. Communism as a hypothesis and political practice is of course much older than the twentieth century and will, in one guise or the other, continue in the future. Excavating the historical-geographical variations and imaginations of the communist invariant requires reexamination and reevaluation. Communism as an idea manifests itself concretely each time people come together in-common, not only to demand equality, to demand their place within the edifice of state and society, but also to stage their capacity for self-organization and self-management, and to enact the democratic promise, thereby changing the frame of what is considered possible and revolutionizing the very parameters of state and government, while putting tentatively and experimentally new organizational forms in their place.

The key task, therefore, is to stop and think, to think through the communist hypothesis and its meaning for a twenty-first-century emancipatory, free, and egalitarian politics. While the pessimism of the intellect over the past few decades, combined with the skepticism of the critical theorist, usurped the will for radical change, the realization of the communist hypothesis requires a new courage of the intellect to break down barriers and taboos, to dare to consider and propose universalizing emancipatory and democratic politics again and to trust the demands formulated by those who have no voice, who have no part; to trust the will for change and to embrace the task of testing radically the truth of the communist hypothesis, a truth that can only be established through a new emancipatory political sequence.

The communist hypothesis forces itself onto the terrain of the political through the process of subjectivation, a coming into being through voluntarist actions, procedures, and performances, of collective embodiments of fidelity to the presumption of equality and freedom. It is a faithful fidelity to the practical possibility of communism, but without ultimate guarantee in history, geography, theory, the party or the state. Communism is an idea without ultimate ground, but with extraordinary emancipatory mileage. The proletarian subjects are those who assemble together, not only to demand freedom and equality but also to take it, to carve out, occupy, and organize the spaces for the enactment of this politics, already experimented with in localized practices of militant groups. The historical-geographical invariant of the communist hypothesis requires serious intellectual engagement in order to tease out what an equal, free, and self-organizing being-in-common for the twenty-first century might be all about. This is a formidable task to be asked of the communist (common) intellect. It will require serious theoretical reconceptualization, a restoration of the trust in our theories, a courageous engagement with painful histories and geographies, and, above all, abandoning the fear of failing again. The fear of failing has become so overwhelming that fear of real change is all that is left; resistance is as far as our horizons reach—transformation, it seems, can no longer be thought, let alone practiced. The injunction scripted by the communist hypothesis is one that urges communist intellectuals to muster the courage to confront the risk of failing again. There is no alternative. We either manage what exists to the best of our humanitarian abilities or think through the possibilities of reimagining and realizing the communist hypothesis for the twenty-first century. This will have to take the form of a "communist geography, a geography of the 'real movement, which abolishes the present state of things'" (Mann 2008, 921).

Notes

Preface

1. I am grateful to the editors and the publishers of the following articles on which parts of the book are based (each relevant chapter is noted in parentheses): "Governance Innovation and the Citizen: The Janus Face of Governance-beyond-the-State," *Urban Studies* 42 (2005): 1991–2006 (chapter 1); "Interrogating Post-Democracy: Reclaiming Egalitarian Political Spaces," *Political Geography* 30 (2011): 370–380 (chapter 2); "Where Is the Political? Insurgent Mobilizations and the Incipient 'Return of the Political,'" *Space and Polity* 18, no. 2 (2014): 122–136 (chapter 3); "Trouble with Nature: 'Ecology as the New Opium for the People,'" in *Conceptual Challenges for Planning Theory*, ed. J. Hillier and P. Healey (Farnham: Ashgate, 2010), 299–320 (chapter 4); "Apocalypse Forever? Post-Political Populism and the Spectre of Climate Change," *Theory, Culture, Society* 27 (2010): 213–232 (chapter 5); "Urbanization and Environmental Futures: Politicizing Urban Political Ecologies," in *Handbook of Political Ecology*, ed. T. Perreault, G. Bridge, and J. McCarthy (London and New York: Routledge, 2015), 609–619 (chapter 6); "Insurgent Architects, Radical Cities and the Promise of the Political," in *The Post-Political and Its Discontents: Spaces of Depoliticization, Specters of Radical Politics*, ed. J. Wilson and E. Swyngedouw (Edinburgh: Edinburgh University Press, 2014), 169–188 (chapter 7); "The Communist Hypothesis and Revolutionary Capitalisms: Exploring the Idea of Communist Geographies for the 21st Century," *Antipode* 41, no. 6 (2010): 1439–1460 (chapter 8).

2 Interrogating Post-Democratization

1. An agonist perspective is skeptical about the political possibilities to transcend or eliminate deep divisions and conflicts within society. Disagreement and tension are an integral part of the social. Chantel Mouffe, for example, distinguishes "between two types of political relations: one of *antagonism* between enemies, and one of *agonism* between adversaries. We could say that the aim of democratic politics is to transform an 'antagonism' into an 'agonism.' ... Far from jeopardizing democracy, agonistic confrontation is in fact its very condition of existence" (Mouffe 1999,

755–756). Agonism departs from the antagonistic friend-enemy position advocated by the Schmittian theorization of the political (see Schmitt 1996), but "identifies the constructive dimension of contestation" (Honig 1993, 15).

3   Theorizing the Political Difference

1. For a critical account of this perspective, see McNay 2014.

2. The philosophical question of the absent ground of the political, the void or excess that institutes the political, opens new perspectives for thinking "the political" again. In the literature, four openings are customarily offered. The "ethical" turn, indebted to Levinas and pursued most consistently by the Heideggerians (see, for example, Lacoue-Labarthe and Nancy 1997; Barnett 2004; or Critchley 2007), insists on how a post-foundational position forces us to take an ethical stance. Claude Lefort's view focuses on how the place of power in democracy is structurally vacant as a site that can be claimed by anyone and anything, at least temporarily (see Lefort 1989; Flynn 2005, and Ingram 2006). The Lacanian psycho-analytically inflected position insists on the irreducible gap between politics as the instituted symbolic order on the one hand and the return of the Real as moment of immanent political opening on the other (Stavrakakis 2006). Finally, there are those who insist on the presumption of equality in the democratic political or, more generally, who start from the partisan position of the possibility of an emancipatory politics. Jacques Rancière, Alain Badiou, Slavoj Žižek, and Etienne Balibar, among others, share this perspective, although they will develop it in different ways. They also share a fundamental distrust of the attempt of the philosophers mentioned earlier to "retreat the political" in a purely philosophical-ethical manner. These authors are also deeply concerned with the political, social, and cultural conditions that characterize the present world, the state of the situation, which they variably describe as post-political or post-democratic. The latter terms evoke respectively a particular tactic of suturing the political, of disavowing the proper political dimension on the one hand and on the other hand, serve as metaphors to capture the institutional forms (including the technologies of governing) that characterize contemporary politics. In addition, the axiomatic invocation of equality as the premise for emancipatory politics distinguishes their position from the ethico-philosophical invocation of an infinite responsibility for the "other" as premise for a living together, a coming community of being-with-others (Barnett 2004). There are of course significant overlaps and common arguments shared by these approaches as well as significant differences. Exploring these is beyond the scope of this book.

3. For an excellent spatialized account, see Gunder and Hillier 2009.

4. I draw here, in particular, on the axiomatic presumption of equality in emancipatory (or democratic) politics (as explored by Badiou and Rancière). For Claude Lefort too, equality is the axiomatically given condition of the democratic invention, which is itself of course historically and geographically contingent. Etienne Balibar also insists that equality is the axiom on which democracy is founded, together with

liberty (what he calls *égaliberté*) (see Dikeç 2003). Badiou is more skeptical about the political uses of "liberty"—he fears this has become too much infected and appropriated by "liberalism" or, in his words, the "capitalo-parliamentary order."

5. The "Organisation Politique" of which Alain Badiou has been a founding and activist member exemplifies such tactics of minimal distance while maintaining an absolute fidelity to the principle of equality as founding gesture (see Hallward 2003a). For Badiou, the passage to the act, the process of political subjectivation emerges on the edges of the void of the state of the situation, the supernumerary part that cannot be fully accounted for within the state of the situation, yet announces the transformation of the state of the situation.

4 Post-Politicizing the Environment

1. The subtitle of this chapter is taken from Žižek 2008a.
2. See http://www.statistics.gov.uk (accessed August 30, 2006).
3. Of course, the geo-philosophical thought of Deleuze and Guattari articulates complexity theory in important ways and has spawned an exciting, albeit occasionally bewildering, literature that takes relationality, indeterminacy, and the radical heterogeneities of natures seriously (see, among others, Conley 1996; Deleuze and Guattari 1994; Herzogenrath 2008; Hillier 2009).
4. This particular semiotic perspective draws on Slavok Žižek's reading of Jacques Lacan's psychoanalytic interpretations of the Imaginary, the Real, and the Symbolic (see Žižek 1989; Lacan 1977, 1993).
5. *Objet petit a* or *objet a* is, for Lacan, the object of desire. It can be any object (separated from the subject) that sets desire in motion, not to achieve immediate enjoyment, but to circle around it. The *objet petit a* is a surplus meaning, a surplus enjoyment, which was inspired by Marx's concept of surplus value. For further details, see Lacan 1977.
6. Political community is built upon some form of gift giving, of give and take, which necessarily means the recognition of the other as part of yourself, as part of your community; it this premise that is now being displaced and annulled.
7. Frederic Neyrat offers therefore the term "Thanatocene" as a more apt signifier tor the Anthropocene (Neyrat 2016). Bonneuil and Fressoz (2017) also mobilize the Thanatocene as a more appropriate name for our epoch, as it is precisely the war machine that drove much of the environmental destruction that gave birth to the notion of the Anthropocene.

5 Hotting Up

1. Of course, the politics of climate science itself should not be ignored, but this is beyond the scope of the present chapter. For a discussion of the politics of climate science, see Demeritt 2001.

8   Exploring the Idea of Emancipatory Geographies for the Twenty-First Century

1. The term "obscure disaster" is taken from Alain Badiou and refers to the disavowed legacy of actually existing, twentieth-century "communisms" by the contemporary Left. Badiou argues that a critical-philosophical engagement with the causes, conditions of possibility, and lessons to be drawn from this "disaster" is urgently required as part of a political project to found a communism for the twenty-first century.
2. See http://ranciere.blogspot.gr/2009/01/on-idea-of-communism.html (accessed July 10, 2017).

References

Abensour, M. 2004. *Democracy against the State*. Cambridge: Polity Press.

Agamben, G. 1998. *Homo Sacer: Sovereign Power and Bare Life*. Stanford, CA: Stanford University Press.

Agamben, G. 2005. *State of Exception*. Chicago: The University of Chicago Press.

Agamben, G. 2006. "Metropolis. Paper presented at Metropoli/Moltitudine Seminario nomade in tre atti. Venice. http://www.generation-online.org/p/fpagamben4.htm (accessed July 20, 2017).

Agamben, G. 2007. *Qu'est-ce qu'un dispositif*. Paris: Rivages Poche.

Agamben, G., A. Badiou, D. Bensaïd, W. Brown, J.-L. Nancy, J. Rancière, K. Ross, and S. Žižek. 2009. *Démocratie, dans quel Etat?* Paris: La Fabrique.

Akkerman, T., M. Hajer, and J. Grin. 2004. "The Interactive State: Democratisation from Above?" *Political Studies* 52:82–95.

Ali, S. H., and R. Keil. 2008. *Networked Disease: Emerging Infections in the Global City*. Malden, MA: Blackwell.

Allmendinger, P., and G. Haughton. 2010. "Spatial Planning, Devolution, and New Planning Spaces." *Environment and Planning. C, Government & Policy* 28 (5): 803–818.

Andreucci, D., M. García Lamarca, J. Wedekind, and E. Swyngedouw. 2017. "'Value Grabbing': A Political Ecology of Rent." *Capitalism, Nature, Socialism*. doi:10.1080/10 455752.2016.1278027.

Angelo, H., and D. Wachsmuth. 2015. "Urbanizing Urban Political Ecology: A Critique of Methodological Cityism." *International Journal of Urban and Regional Research* 39:16–27.

Arendt, H. 1973. *The Origins of Totalitarianism*. New York: Harcourt Publishers Ltd College Publishers.

Armiero, M. 2014. "Garbage under the Volcano: The Waste Crisis in Campania and the Struggles for Environmental Justice." In *A History of Environmentalism: Local Struggles, Global Histories*, ed. M. Armiero and L. Sedrez, 167–183. London: Bloomsbury.

Badiou, A. 1998. *L'Organisation Politique* 28:2.

Badiou, A. 1999. "Nicolas Poirier: Entretien avec Alain Badiou." *Le Philosophoire* 9:11–25.

Badiou, A. 2005a. *Being and Event*. London: Continuum.

Badiou, A. 2005b. "Politics: A Non-Expressive Dialectics." Paper presented at the conference Is The Politics of Truth Still Thinkable? A conference organized by Slavoj Zizek and Costas Douzinas. Birkbeck Institute for the Humanities, Birkbeck College, London, November 25–26.

Badiou, A. 2008a. "The Communist Hypothesis." *New Left Review* 49:29–42.

Badiou, A. 2008b. "Live Badiou: Interview with Alain Badiou, Paris, December 2007." In *Alain Badiou: Live Theory*, ed. O. Feltham, 136–139. London: Continuum.

Badiou, A. 2008c. *The Meaning of Sarkozy*. London: Verso.

Badiou, A. 2010. *The Communist Hypothesis*. London: Verso.

Badiou, A. 2012. *The Rebirth of History: Times of Riots and Uprisings*. London: Verso.

Badiou, A. 2017. *Que Signifie "Changer Le Monde" 2010–2012*. Paris: Fayard.

Badiou, A., and S. Žižek. 2010. *L'Idée du Communisme*. Paris: Nouvelles Editions Lignes.

Baeten, G. 2009. "Regenerating the South Bank. Reworking the Community and the Emergence of Post-Political Regeneration." In *Regenerating London. Governance, Sustainability and Community in the Global City*, ed. R. Imrie, L. Lees, and M. Raco, 237–253. London: Routledge.

Baeten, G. 2010. "Reactionary Responses to the Environmental Crisis." Paper presented at the International Conference on Environmental Conflicts and Justice. The Institute of Environmental Science and Technology, Autonomous University of Barcelona, July 2–3.

Bakker, K. 2003. *An Uncooperative Commodity: Privatizing Water in England and Wales*. New York; Oxford: Oxford University Press.

Balibar, E. 1993. *Masses, Classes, Ideas: Studies on Politics and Philosophy before and after Marx*. London: Routledge.

Balibar, E. 2010. *La Proposition de l'Egaliberte*. Paris: Presses Universitaires de France.

Barnett, C. 2004. "Deconstructing Radical Democracy: Articulation, Representation and Being-with-Others." *Political Geography* 23:503–528.

Bassett, K. 2008. "Thinking the Event: Badiou's Philosophy of the Event and the Example of the Paris Commune." *Environment and Planning. D, Society & Space* 26 (5): 895–910.

Bassett, K. 2016. "Event, Politics, and Space: Rancière or Badiou?" *Space and Polity* 20:280–293.

Beck, U. 1997. *The Reinvention of Politics: Rethinking Modernity in the Global Social Order.* Cambridge: Polity Press.

Beck, U. 1999. *World Risk Society.* Cambridge: Polity Press.

Bellamy, R., and A. Warleigh. 2001. *Citizenship and Governance in the European Union.* London: Continuum Publishers.

Blühdorn, I. 2006. "Billich will Ich: Post-demokratische Wende und Simulative Demokratie." *Forschungsjournal NSB* 19:72–83.

Blühdorn, I. 2013. "The Governance of Unsustainability: Ecology and Democracy after the Post-Democratic Turn." *Environmental Politics* 22:16–36.

Boltanski, L., and E. Chiapello. 2007. *The New Spirit of Capitalism.* London: Verso.

Bonneuil, C., and J.-B. Fressoz. 2017. *The Shock of the Anthropocene: The Earth, History and Us.* London: Verso.

Bosteels, B. 2005. "The Speculative Left." *South Atlantic Quarterly* 104:751–767.

Bosteels, B. 2011. *The Actuality of Communism.* London: Verso.

Bosteels, B. 2014. "Archipolitics, Parapolitics, Metapolitics." In *Jacques Rancière: Key Concepts*, ed. J. P. Deranty, 80–94. London: Routledge.

Bourdieu, P. 2002. "Against the Policy of Depoliticization." *Studies in Political Economy* 69:31–41.

Bouzarovski, S. 2104. "Energy Poverty in the European Union: Landscapes of Vulnerability." *WIREs Energy and Environment* 3:276–289.

Braun, B. 2006. "Environmental Issues: Global Natures in the Space of Assemblage." *Progress in Human Geography* 30:644–654.

Braun, B. 2015. "New Materialisms and Neoliberal Natures." *Antipode* 47:1–14.

Brechin, G. 2001. *Imperial San Francisco: Urban Power, Earthly Ruin.* Berkeley: University of California Press.

Brenner, N. 2004. *New State Spaces: Urban Governance and the Rescaling of Statehood.* Oxford: Oxford University Press.

Brenner, N., B. Jessop, M. Jones, and G. MacLeod. 2003. "Introduction: State Space in Question." In *State/Space*, ed. N. Brenner, B. Jessop, M. Jones and G. MacLeod, 1–26. Oxford: Blackwell.

Brenner, N., and N. Theodore. 2002. *Spaces of Neoliberalism: Urban Restructuring in North America and Western Europe*. Oxford: Blackwell.

Brossat, A. 2003. *La Démocratie Immunitaire*. Paris: La Dispute.

Brown, W. 2005. *Edgework: Critical Essays on Knowledge and Politics*. Princeton: Princeton University Press.

Brown, W. 2015. *Undoing the Demos: Neoliberalism's Stealth Revolution*. Cambridge, MA: MIT Press.

Buck, D. 2007. "The Ecological Question: Can Capitalism Survive?" In *Socialist Register 2007: Coming to Terms with Nature*, ed. I. Panitch and C. Leys, 60–71. New York: Monthly Review Press.

Buerk, R. 2006. *Breaking Ships: How Supertankers and Cargo Ships Are Dismantled on the Beaches of Bangladesh*. New York: Chamberlain Books.

Bulkeley, H., and M. Betsill. 2005. *Cities and Climate Change: Urban Sustainability and Global Environmental Governance*. London: Routledge.

Bullard, R. D. 1990. *Dumping in Dixie: Race, Class, and Environmental Quality*. Boulder, CO: Westview.

Bumpus, A. G., and D. Liverman. 2008. "Accumulation by Decarbonization and the Governance of Carbon Offsets." *Economic Geography* 84:127–155.

Burchell, G. 1993. "Liberal Government and Techniques of the Self." *Economy and Society* 22:267–282.

Burchell, G. 1996. "Liberal Government and Techniques of the Self." In *Foucault and Political Reason: Liberalism, Neo-Liberalism, and Rationalities of Government*, ed. A. Barry, T. Osborne and N. Rose, 19–36. London: University College London Press.

Campbell, D. 2009. "Move Over Jacko, Idea of Communism Is Hottest Ticket in Town This Weekend." *The Guardian*, March 12.

Canovan, M. 1999. "Trust the People! Populism and the Two Faces of Democracy." *Political Studies* 47:2–16.

Canovan, M. 2005. *The People*. Cambridge: Polity Press.

Caprotti, F. 2014. "Eco-Urbanism and the Eco-City, or Denying the Right to the City?" *Antipode* 46:1285–1303.

Carothers, T., W. Brandt, and M. K. Al-Sayyid. 2000. "Civil Society." *Foreign Policy* 117:18–29.

Castel, R. 1991. "From Dangerous to Risk." In *The Foucault Effect: Studies in Governmentality*, ed. G. Burchell, C. Gordon, and P. Miller, 281–298. Chicago: The University of Chicago Press.

Castells, M. 1993. *The City and the Grassroots: A Cross-Cultural Theory of Urban Social Movements*. Berkeley: University of California Press.

Castoriadis, C. 1987. *The Imaginary Institution of Society*. Cambridge: Polity Press.

Castree, N. 2003. "Environmental Issues: Relational Ontologies and Hybrid Politics." *Progress in Human Geography* 27:203–211.

Castree, N. 2009. "The Environmental Wedge: Neoliberalism, Democracy and the Prospect for a New British Left." In *Feelbad Britain: How to Make It Better*, ed. P. Devine, A. Pearman, and D. Purdy, 222–233. London: Lawrence and Wishart.

Castree, N. 2014. "Geography and the Anthropocene II: Current Contributions." *Geography Compass* 8:450–463.

Castree, N. 2015. "Changing the Anthropo(s)cene: Geographers, Global Environmental Change and the Politics of Knowledge." *Dialogues in Human Geography* 5:301–316.

Coaffee, J., and P. Healey. 2003. "'My Voice: My Place': Tracking Transformations in Urban Governance." *Urban Studies* 40:1979–1999.

Cochrane, A. 2010. "Exploring the Regional Politics of 'Sustainability': Making up Sustainable Communities in the South-East of England." *Environmental Policy and Governance* 20:370–381.

Coe, N., P. F. Kelly, and H. Yeung. 2007. *Economic Geography: A Contemporary Introduction*. Oxford: Blackwell.

Conley, V. 1996. *Ecopolitics: The Environment in Poststructural Thought*. London: Routledge.

Cook, I., and E. Swyngedouw. 2012. "Cities, Social Cohesion and the Environment: Towards a Future Research Agenda." *Urban Studies* 49:1938–1958.

Cooke, B., and U. Kothari. 2001. *Participation: The New Tyranny?* London: Zed Books.

Critchley, S. 2007. *Ethics of Commitment, Politics of Resistance*. London: Verso.

Cronon, W. 1991. *Nature's Metropolis: Chicago and the Great West*. New York: W. W. Norton and Company.

Crouch, C. 2000. *Coping with Post-Democracy*. Fabian Ideas 598. London: The Fabian Society.

Crouch, C. 2004. *Post-Democracy*. Cambridge: Polity Press.

Crouch, C. 2015. "The March towards Post-Democracy, Ten Years On." *Political Quarterly* 87:71–75.

Cruikshank, B. 1993. "Revolutions within: Self-Governance and Self-Esteem." *Economy and Society* 22:327–344.

Cruikshank, B. 1994. "The Will to Empower: Technologies of Citizenship and the War on Poverty." *Socialist Review* 23:29–55.

Crutzen, P. J., and E. F. Stoermer. 2000. "The 'Anthropocene.'" *Global Change Newsletter* 41:17–18.

Davis, M. 1998. *Ecology of Fear: Los Angeles and the Imagination of Disaster*. New York: Metropolitan Books.

Davis, M. 2002. *Dead Cities*. New York: The New Press.

Davis, O. 2010. *Jacques Rancière*. Cambridge: Polity Press.

De Cauter, L. 2004. *The Capsular Civilization: On the City in the Age of Fear*. Rotterdam: NAI Publishers.

Dean, J. 2006. *Žižek's Politics*. New York: Routledge.

Dean, J. 2009. *Democracy and Other Neoliberal Fantasies: Communicative Capitalism and Left Politics*. Durham, NC: Duke University Press.

Dean, J. 2012. *The Communist Horizon*. London: Verso.

Dean, M. 1995. "Governing the Unemployed Self in an Active Society." *Economy and Society* 24:559–583.

Dean, M. 1999. *Governmentality: Power and Rule in Moderns Society*. London: Sage.

Debord, G. 1967. *La Société du Spectacle*. Paris: Buchet-Chastel.

Debord, G. 1994. *The Society of the Spectacle*. New York: Zone.

Deleuze, G., and F. Guattari. 1994. *What Is Philosophy?* New York: Columbia University Press.

Demeritt, D. 2001. "Being Constructive about Nature." In *Social Nature: Theory, Practice and Politics*, ed. B. Braun and N. Castree, 22–40. Oxford: Blackwell.

Derrida, J. 1982. "Of an Apocalyptic Tone Recently Adopted in Philosophy." *Semieia* 23:63–97.

Dikeç, M. 2001. "Justice and the Spatial Imagination." *Environment & Planning A* 33:1785–1805.

Dikeç, M. 2003. "Police, Politics, and the Right to the City." *GeoJournal* 58:91–98.

Dikeç, M. 2005. "Space, Politics and the Political." *Environment and Planning. D, Society & Space* 23:171–188.

Dikeç, M. 2007. *Badlands of the Republic: Space, Politics and French Urban Policy.* Oxford: Blackwell.

Dikeç, M. 2015. *Space, Politics and Aesthetics.* Edinburgh: Edinburgh University Press.

Dikeç, M. 2017. *Urban Rage: The Revolt of the Excluded.* New Haven: Yale University Press.

Dikeç, M., and E. Swyngedouw. 2017. "Theorizing the Politicizing City." *International Journal of Urban and Regional Research* 41:1–18. doi:10.1111/1468-2427.12388.

Diken, B. 2004. "From Refugee Camps to Gated Communities: Biopolitics and the End of the City." *Citizenship Studies* 8:83–106.

Diken, B., and C. Laustsen. 2004. "7/11, 9/11, and Post-Politics." Lancaster: Department of Sociology, Lancaster University, UK. http://www.lancaster.ac.uk/fass/resources/sociology-online-papers/papers/diken-laustsen-7-11-9-11-post-politics.pdf (accessed January 8, 2018).

Dingwerth, K. 2004. "Democratic Governance beyond the State: Operationalising an Idea." Global Governance Working Paper 14. Amsterdam: The Global Governance Project; www.glogov.org (accessed May 16, 2005).

Docherty, I., R. Goodlad, and R. Paddison. 2001. "Civic Culture, Community and Citizen Participation in Contrasting Neighbourhoods." *Urban Studies* 38:2225–2250.

Donzelot, J. 1984. *L'Invention du Social: Essai sur le Déclin des Passions Politiques.* Paris: Seuil.

Donzelot, J. 1991. "The Mobilization of Society." In *The Foucault Effect: Studies in Governmentality*, ed. G. Burchell, C. Gordon, and P. Miller, 169–179. Chicago: University of Chicago Press.

Douzinas, C., and S. Žižek. 2010. *The Idea of Communism.* London: Verso.

Dryzek, J. 2000. *Deliberative Democracy and Beyond.* Oxford: Oxford University Press.

Edwards, M. 2002. "Herding Cats? Civil Society and Global Governance." *New Economy* 9:71–76.

Ekers, M., G. Hart, S. Kipfer, and A. Loftus. 2013. *Gramsci: Space, Nature, Politics.* Oxford: Wiley-Blackwell.

Elden, S. 2007. "Governmentality, Calculation, Territory." *Environment and Planning. D, Society & Space* 25:562–580.

Ernstson, H., and E. Swyngedouw. 2018. "Interrupting the Anthropo-Obscene: Immuno-Biopolitics and the Re-Invention of the Political in the Anthropocene."

Forthcoming in *Urban Political Ecology in the Anthropo-Obscene: Interruptions and Possibilities*, ed. H. Ernstson and E. Swyngedouw. London: Routledge.

Esposito, R. 2008. *Bios: Biopolitics and Philosophy*. Minneapolis: University of Minnesota Press.

Esposito, R. 2011. *Immunitas*. Cambridge: Polity Press.

European Commission. 2001. "European Governance: A White Paper." COM(2001) 428 final. Brussels: Commission of the European Communities.

Featherstone, D. 2008. *Resistance, Space and Political Identities: The Making of Counter-Global Networks*. Oxford: Wiley-Blackwell.

Flynn, B. 2005. *The Philosophy of Claude Lefort: Interpreting the Political*. Evanston: Northwestern University Press.

Foucault, M. 1979. "On Governmentality." *Ideology & Consciousness* 6:5–21.

Foucault, M. 1982. "The Subject and Power." In *Michel Foucault: Beyond Structuralism and Hermeneutics*, ed. H. Dreyfus and P. Rabinow, 208–226. Brighton: Harvester.

Foucault, M. 1984. "Space, Knowledge, and Power." In *The Foucault Reader*, ed. P. Rabinow, 239–256. Harmondsworth: Penguin Books.

Foucault, M. 1991. "Governmentality." In *The Foucault Effect: Studies in Governmentality*, ed. G. Burchell, C. Gordon and P. Milller, 87–104. Hempstead: Harverster Wheatsheaf.

Foucault, M. 2004a. *Naissance de la Biopolitique: Cours au Collège de France (1978–1979)*. Paris: Ed. Seuil.

Foucault, M. 2004b. *Sécurité, Territoire, Population: Cours au Collège de France (1977–1978)*. Paris: Ed. Seuil.

Foucault, M. 2007. *Territory, Population: Lectures at the Collège de France 1977–1978*. London: Palgrave Macmillan.

Freidberg, S. 2004. *French Beans and Food Scares: Culture and Commerce in an Anxious Age*. Oxford: Oxford University Press.

Gandy, M. 2002. *Concrete and Clay: Reworking Nature in New York City*. Cambridge, MA: MIT Press.

Gandy, M. 2005. "Cyborg Urbanization: Complexity and Monstrosity in the Contemporary City." *International Journal of Urban and Regional Research* 29:26–49.

Gandy, M. 2006. "Urban Nature and the Ecological Imaginary." In *In the Nature of Cities: Urban Political Ecology and the Metabolism of Urban Environments*, ed. N. Heynen, M. Kaika, and E. Swyngedouw, 62–72. London: Routledge.

García Lamarca, M. 2017. "Creating Political Subjects: Collective Knowledge and Action to Enact Housing Rights in Spain." *Community Development Journal* 52 (3): 421–435.

García Lamarca, M. 2017. "From Occupying Plazas to Recuperating Housing: Insurgent Practices in Spain." *International Journal of Urban and Regional Research* 41:37–53.

García Lamarca, M., and M. Kaika. 2016. "'Mortgaged Lives': The Biopolitics of Debt and Housing Financialisation." *Transactions of the Institute of British Geographers* 41:313–327. doi:10.1111/tran.12126.

Garcia, P.-O. 2015. "Sous l'Adaptation, l'Immunité. Etude sur le Discours de l'Adaptation au Changement Climatique." PhD dissertation, Université Grenoble Alpes, Grenoble.

Getimis, P., and G. Kafkalas. 2002. "Comparative Analysis of Policy-Making and Empirical Evidence on the Pursuit of Innovation in Sustainability." In *Participatory Governance in Multi-Level Context: Concepts and Experience*, ed. P. Getimis, H. Heinelt, G. Kafkalas, R. Smith, and E. Swyngedouw, 107–131. Opladen: Leske & Budrich.

Gibbs, D., and R. Krueger. 2007. *The Sustainable Development Paradox*. New York: Guilford Press.

Giddens, A. 2009. *The Politics of Climate Change*. Cambridge: Polity Press.

Giroux, H. A. 2004. *The Terror of Neoliberalism: Authoritarianism and the Eclipse of Democracy*. Boulder, CO: Paradigm Publishers.

Glassman, J. 2007. "Post-Democracy." *Environment & Planning A* 39:2037–2042.

González, S., and P. Healey. 2005. "A Sociological Institutionalist Approach to the Study of Innovation in Governance Capacity." *Urban Studies* 42:2055–2069.

Goonewardena, K., and K. N. Rankin. 2004. "The Desire Called Civil Society: A Contribution to the Critique of a Bourgeois Category." *Planning Theory* 3:117–149.

Gordon, C. 1991. "Governmental Rationality." In *The Foucault Effect: Studies in Governmentality*, ed. G. Burchell, C. Gordon, and P. Miller, 1 51. Chicago: The University of Chicago Press.

Gould, S. J. 1980. *The Panda's Thumb*. New York: W. W. Norton.

Graham, S. 2009. *Cities under Siege: The New Military Urbanism*. London: Verso.

Graham, S., and S. Marvin. 2001. *Splintering Urbanism*. London: Routledge.

Gramsci, A. 1971. *Selections from the Prison Notebooks*. London: Lawrence & Wishart.

Grote, J., and B. Gbikpi. 2002. *Participatory Governance: Societal and Political Implications*. Opladen: Leske and Budrich.

Gunder, M., and J. Hillier. 2009. *Planning in Ten Words or Less: A Lacanian Entanglement with Spatial Planning*. Farnham: Ashgate.

Hajer, M. 2003a. *Deliberative Policy Analysis: Understanding Governance in the Network Society*. Cambridge: University Press.

Hajer, M. 2003b. "Policy without Polity? Policy Analysis and the Institutional Void." *Policy Sciences* 36:175–195.

Hallward, P. 2003a. *Badiou: A Subject to Truth*. Minneapolis: University of Minnesota Press.

Hallward, P. 2003b. "Introduction. Jacques Rancière—Politics and Aesthetics. An Interview." *Angelaki* 8 (2):191–193.

Hallward, P. 2005. "Jacques Rancière and the Subversion of Mastery." *Paragraph* 28:26–45.

Hallward, P. 2009. "The Will of the People: Notes towards a Dialectical Voluntarism." *Radical Philosophy* 155:17–29.

Hamel, P. 2002. "Conclusion: Enjeux Institutionnels et Défis Politiques." In *Développement Durable et Participation Publique: De la Contestation Ecologique aux Défis de la Gouvernance*, ed. C. Gendron and J. C. Vaillancourt, 377–392. Montreal: Presses Universitaires de Montréal.

Hamilton, C. 2013. *Earthmasters: The Dawn of the Age of Climate Engineering*. London: Yale University Press.

Haraway, D. 1991. *Simians, Cyborgs, and Women: The Reinvention of Nature*. London: Free Association Books.

Hardt, M. 2010. "The Common in Communism." *Rethinking Marxism: A Journal of Economics, Culture & Society* 22:346–356.

Hardt, M., and A. Negri. 2001. *Empire*. Cambridge, MA: Harvard University Press.

Hardt, M., and A. Negri. 2004. *Multitude*. London: Penguin Books.

Hardt, M., and A. Negri. 2011. "The Fight for 'Real Democracy' at the Heart of Occupy Wall Street." *Foreign Affairs*, October 11. https://www.foreignaffairs.com/articles/north-america/2011-10-11/fight-real-democracy-heart-occupy-wall-street (accessed July 12, 2015).

Harvey, D. 1993. "The Nature of Environment: Dialectics of Social and Environmental Change." In *Real Problems, False Solutions*, ed. R. Milliband and L. Panitch, 1–51. London: Merlin Press.

Harvey, D. 1996. *Justice, Nature and the Geography of Difference*. Oxford: Blackwell.

Harvey, D. 2005. *Neoliberalism: A Short History*. Oxford: University Press.

Harvey, D. 2009. "Is This Really the End of Neoliberalism?" http://www.counter-punch.org/harvey03132009.html (accessed July 21, 2015).

Harvey, D. 2012. *Rebel Cities: From the Right to the City to the Urban Revolution.* London: Verso.

Henderson, G. 2009. "Marxist Political Economy and the Environment." In *A Companion to Environmental Geography,* ed. N. Castree, D. Demeritt, D. Liverman, and B. Rhoads, 266–293. Oxford: Wiley-Blackwell.

Hermet, G. 2009. *L'Hiver de la Démocratie ou le Nouveau Régime.* Paris: Armand Colin.

Hertz, N. 2002. *The Silent Takeover: Global Capitalism and the Death of Democracy.* London: Arrow.

Herzogenrath, B. 2008. *An (Un)Likely Alliance: Thinking Environment(s) with Deleuze/Guattari.* Newcastle upon Tyne, UK: Cambridge Scholars Publishing.

Heynen, N., M. Kaika, and E. Swyngedouw. 2006. *In the Nature of Cities: Urban Political Ecology and the Metabolism of Urban Environments.* London: Routledge.

Heynen, N., H. E. Kurtz, and A. Trauger. 2012. "Food Justice, Hunger and the City." *Geography Compass* 6:304–311.

Hewlett, N. 2007. *Badiou, Balibar, Rancière: Re-Thinking Emancipation.* London: Continuum.

Hillier, J. 2009. "Assemblages of Justice: The 'Ghost Ships' of Graythorp." *International Journal of Urban and Regional Research* 33:640–661.

Hirst, P. 1995. "Can Secondary Associations Enhance Democratic Governance?" In *Associations and Democracy,* ed. E. O. Wright, 101–114. London: Verso.

Holloway, J. 2002. *Change the World without Taking Power: The Meaning of Revolution Today.* London: Pluto.

Honig, B. 1993. *Political Theory and the Displacement of Politics.* Ithaca: Cornell University Press.

Hornborg, A. 2009. "Zero-Sum World: Challenges in Conceptualizing Environmental Load Displacement and Ecologically Unequal Exchange in the World-System." *International Journal of Comparative Sociology* 50 (3–4): 237–262.

Hulme, M. 2008. "Geographical Work at the Boundaries of Climate Change." *Transactions of the Institute of British Geographers* 33:5–11.

Hulme, M. 2009. *Why We Disagree about Climate Change.* Cambridge: Cambridge University Press.

Ingram, J. D. 2006. "The Politics of Claude Lefort's Political: Between Liberalism and Radical Democracy." *Thesis Eleven* 87:33–50.

Intergovernmental Panel for Climate Change. 2007. *Fourth Assessment Report, Climate Change 2007*. Cambridge: Cambridge University Press.

Intergovernmental Panel for Climate Change. 2009. *Climate Change 2007: Impacts, Adaptation and Vulnerability. Working Group II Contribution to the Fourth Assessment Report of the IPCC*. Cambridge: Cambridge University Press.

Intergovernmental Panel for Climate Change. 2013. "Summary for Policymakers." In *Climate Change 2013: The Physical Science Basis. Contribution of Working Group I to the Fifth Assessment Report of the Intergovernmental Panel on Climate Change*, ed. T. F. Stocker, D. Qin, G.-K. Plattner, M. Tignor, S. K. Allen, J. Boschung, A. Nauels, Y. Xia, V. Bex, and P. M. Midgley, 3–29. Cambridge: Cambridge University Press.

Invisible Committee, The. 2009. *The Coming Insurrection*. Cambridge, MA: MIT Press.

Isin, E. F. 2000. *Democracy, Citizenship and the Global City*. London: Routledge.

Isin, E. F. 2002. *Being Political: Genealogies of Citizenship*. Minneapolis: University of Minnesota Press.

Jameson, F. 2003. "Future City." *New Left Review* 21:65–79.

Jay, M. 1994. "The Apocalyptic Imagination and the Inability to Mourn." In *Rethinking Imagination: Culture and Creativity*, ed. G. Robinson and J. Rundell, 30–47. New York: Routledge.

Jessop, B. 1995. "The Regulation Approach: Governance and Post-Fordism: Alternative Perspectives on Economic and Political Change." *Economy and Society* 24: 307–333.

Jessop, B. 1998. "The Rise of Governance and the Risks of Failure: The Case of Economic Development." *International Social Science Journal* 50 (155): 29–45.

Jessop, B. 2002a. *The Future of the Capitalist State*. Cambridge: Polity Press.

Jessop, B. 2002b. "Governance and Meta-Governance: On Reflexivity, Requisite Variety, and Requisite Irony." In *Governance, Governmentality and Democracy*, ed. H. Bang, 142–172. Manchester: Manchester University Press.

Jessop, B. 2002c. "Liberalism, Neoliberalism and Urban Governance: A State-Theoretical Perspective." *Antipode* 34:452–472.

Jessop, B. 2005. "Political Ontology, Political Theory, Political Philosophy, and the Ironic Contingencies of Political Life." In *Der Mensch—ein Zoon Politikon*, ed. H. Schmidinger and C. Sedmak, 189–208. Darmstadt: Wissenschaftliche Buchgesellschaft.

Jörke, D. 2005. "Auf dem Weg zur Postdemokratie." *Leviathan* 33:482–491.

Jörke, D. 2008. "Postdemocracia en América Latina y Europa Revista de Sociología de la Universidad de Chile." *Revista de Sociología de la Universidad de Chile* 21:124–135.

Kaika, M. 2005. *City of Flows: Nature, Modernity and the City.* New York: Routledge.

Kaika, M. 2010. "Architecture and Crisis: Re-Inventing the Icon, Re-Imag(in)ing London and Re-branding the City." *Transactions of the Institute of British Geographers* 35:453–474.

Kaika, M. 2011. "Autistic Architecture: The Fall of the Icon and the Rise of the Serial Object of Architecture." *Environment and Planning. D, Society & Space* 29:968–992.

Kaika, M. 2017. "Between Compassion and Racism: How the Biopolitics of Neoliberal Welfare Turns Citizens into Affective 'Idiots.'" *European Planning Studies* 25:1275–1291.

Kaika, M. 2018. "Between Racism and Philanthropy: The Biopolitics of Affect … or … How to Turn Citizens into 'Idiots' through Privatized Welfare and Household Debt." Forthcoming in *Urban Political Ecology in the Anthropo-Obscene: Interruptions and Possibilities*, ed. Henrik Ernstson and Erik Swyngedouw. London: Routledge.

Kakigianno, M., and J. Rancière. 2013. "A Precarious Dialogue." *Radical Philosophy* 178:18–25.

Katz, C. 1995. "Under the Falling Sky: Apocalyptic Environmentalism and the Production of Nature." In *Marxism in the Postmodern Age*, ed. A. Callari, S. Cullenberg, and C. Biewener, 276–282. New York: Guilford Press.

Kaulinfreks, F. 2008. "'Fuck Normalization'—Young Urban 'Troublemakers' as Meaningful Political Actors." *Resistance Studies Magazine* 3:35–51.

Keil, R. 2003. "Urban Political Ecology.' *Urban Geography* 24:723–738.

Keil, R. 2005. "Progress Report: Urban Political Ecology." *Urban Geography* 26:640–651.

Kenis, A., and E. Mathijs. 2014. "Climate Change and Post-Politics: Repoliticizing the Present by Imagining the Future?" *Geoforum* 52:148–156.

Kooiman, J. 1995. *Modern Governance: Levels, Models and Orders of Socio-Political Interaction.* London: Sage.

Kooiman, J. 2000. "Societal Governance: Levels, Models and Orders of Socio-Political Interaction." In *Debating Governance: Authority, Steering and Democracy*, ed. J. Pierre, 138–166. Oxford: Oxford University Press.

Kooiman, J. 2003. *Governing as Governance.* London: Sage.

Krippner, G. R. 2005. "The Financialization of the American Economy." *Socio-Economic Review* 3:173–208.

Lacan, J. 1977. *The Seminar of Jacques Lacan, Book XI: The Four Fundamental Concepts of Psychoanalysis.* London: Hogarth Press.

Lacan, J. 1993. *The Seminar of Jacques Lacan, Book II: The Psychoses 1955–1956.* New York: W. W. Norton.

Laclau, E. 2005. *On Populist Reason.* London: Verso.

Laclau, E., and C. Mouffe. 1985. *Hegemony and Socialist Strategy: Towards a Radical Democratic Politics.* London: Verso.

Lacoue-Labarthe, P., and J.-L. Nancy. 1997. *Retreating the Political.* London: Routledge.

Latour, B. 1993. *We Have Never Been Modern.* New York; London: Harvester Wheatsheaf.

Latour, B. 2004. *Politics of Nature: How to Bring the Sciences into Democracy.* Cambridge, MA: Harvard University Press.

Latour, B. 2005. *Reassembling the Social: An Introduction to Actor-Network-Theory.* Oxford: Oxford University Press.

Le Galès, P. 1995. "Du Gouvernement Local à la Gouvernance Urbaine." *Revue Francaise de Science Politique* 45:57–95.

Le Galès, P. 2002. *European Cities: Social Conflicts and Governance.* Oxford: Oxford University Press.

Lefort, C. 1986. *The Political Forms of Modern Society: Bureaucracy, Democracy, Totalitarianism.* Cambridge: Polity Press.

Lefort, C. 1989. *Democracy and Political Theory.* Minneapolis: University of Minnesota Press.

Lefort, C. 1994. *L'Invention Démocratique, Les Limites de la Domination Totalitaire.* Paris: Fayard.

Lemke, T. 2001. "'The Birth of Bio-Politics': Michel Foucault's Lectures at the College de France on Neo-Liberal Governmentality." *Economy and Society* 30:190–207.

Lemke, T. 2002. "Foucault, Governmentality, and Critique." *Rethinking Marxism* 14:49–64.

Levene, M. 2005. "The Reality and Urgency of Human-created Climate Change." *Rescue!History: A Manifesto for the Humanities in the Age of Climate Change—An Appeal for Collaborators.* http://www.rescue-history.org.uk (accessed August 2, 2017).

Levins, R., and R. Lewontin. 1985. *The Dialectical Biologist.* Cambridge, MA: Harvard University Press.

Lévy, J., J. Rennes, and D. Zerbib. 2007. "Jacques Rancière: 'Les Territoires de la Pensée Partagée.'" In *EspaceTemps.net, Actuel.* http://espacestemps.net/document2142.html (accessed July 12, 2016).

Lewontin, R. C., and R. Levins. 2007. *Biology under the Influence: Dialectical Essays on Ecology, Agriculture, and Health*. New York: Monthly Review Press.

Linhardt, D., and F. Muniesa. 2011. "Tenir Lieu de Politique: Le Paradoxe des '"Politiques d'Economisation."' *Politix* 95:6–21.

Liverman, D. 2009. "Conventions of Climate Change: Constructions of Danger and the Dispossession of the Atmosphere." *Journal of Historical Geography* 35:279–296.

Lovelock, J. 2010. *The Vanishing Face of Gaia: A Final Warning*. Harlow: Penguin.

Low, S., and N. Smith. 2005. *The Politics of Public Space*. New York: Routledge.

MacLeod, G. 1999. "Entrepreneurial Spaces, Hegemony, and State Strategy: The Political Shaping of Privatism in Lowland Scotland." *Environment & Planning A* 31:345–375.

MacLeod, G. 2011. "Urban Politics Reconsidered: Growth Machine to Post-Democratic City?" *Urban Studies* 48:2629–2660.

Mann, G. 2008. "A Negative Geography of Necessity." *Antipode* 40:921–934.

March, H., and D. Saurí. 2013. "The Unintended Consequences of Ecological Modernization: Debt-Induced Reconfiguration of the Water Cycle in Barcelona." *Environment & Planning A* 45:2064–2083.

Marchart, O. 2007. *Post-Foundational Political Thought*. Edinburgh: Edinburgh University Press.

Marchart, O. 2011. "Democracy and Minimal Politics: The Political Difference and Its Consequences." *South Atlantic Quarterly* 110:965–973.

Marquand, D. 2004. *Decline of the Public: The Hollowing Out of Citizenship*. Cambridge: Polity Press.

Marvin, S., and W. Medd. 2006. "Metabolisms of Obecity: Flows of Fat through Bodies, Cities, and Sewers." *Environment & Planning A* 38:313–324.

Marx, K. 1967. *Economic and Philosophical Manuscripts of 1844*. Moscow: Progress Publishers.

Marx, K., and F. Engels. 1987. *The German Ideology: Introduction to a Critique of Political Economy*. London: Lawrence and Wishart Ltd.

Masjuan, E., H. March, E. Domene, and D. Saurí. 2008. "Conflicts and Struggles over Urban Water Cycles: The Case of Barcelona 1880–2004." *Tijdschrift voor Economische en Sociale Geografie* 99:426–439.

May, T. 2008. *The Political Thought of Jacques Rancière: Creating Equality*. Edinburgh: Edinburgh University Press.

May, T. 2010. *Contemporary Political Movements and the Thought of Jacques Ranciere: Equality in Action*. Edinburgh: Edinburgh University Press.

Mayer, M., C. Thörn, and H. Thörn. 2016. *Urban Uprisings: Challenging the Neoliberal City in Europe*. London: Palgrave.

McEwan, I. 2010. *Solar*. London: Jonathan Cape.

McNay, L. 2014. *The Misguided Search for the Political*. Cambridge: Polity Press.

Meadows, D. H., D. L. Meadows, J. Randers, and W. W. Behrens III. 1972. *The Limits to Growth: A Report for the Club of Rome's Project on the Predicament of Mankind*. New York: Universe Books.

Merrifield, A. 2011. *Magical Marxism: Subversive Politics and the Imagination*. London: Pluto Press.

Merrifield, A. 2013. *The Politics of the Encounter: Urban Theory and Protest under Planetary Urbanization*. Athens, GA: University of Georgia Press.

Miller, P. 1992. "Accounting and Objectivity: The Invention of Calculating Selves and Calculable Spaces." *Annals of Scholarship* 9:61–86.

Minca, C. 2005. "The Return of the Camp." *Progress in Human Geography* 29:405–412.

Mitchell, K. 2002. "Transnationalism, Neoliberalism and the Rise of the Shadow State." *Economy and Society* 30:165–189.

Monstadt, J. 2009. "Conceptualizing the Political Ecology of Urban Infrastructures: Insights from Technology and Urban Studies." *Environment & Planning A* 41:1924–1942.

Morgan, B. 2003. "The Economization of Politics: Meta-Regulation as a Form of Non-Judicial Legality." *Social & Legal Studies* 12:489–523.

Morton, T. 2007. *Ecology without Nature: Rethinking Environmental Aesthetics*. Cambridge, MA: Harvard University Press.

Mostafavi, M., and G. Doherty. 2010. *Ecological Urbanism*. Zurich: Lars Müller.

Mouffe, C. 1999. "Deliberative Democracy or Agonistic Pluralism?" *Social Research* 66:745–758.

Mouffe, C. 2000. *The Democratic Paradox*. London: Verso.

Mouffe, C. 2005. *On the Political. Thinking in Action*. London: Routledge.

Mouffe, C. 2013. *Agonistics: Thinking the World Politically*. London: Verso.

Moulaert, F., F. Martinelli, and E. Swyngedouw. 2006. "Towards Alternative Model(s) of Local Innovation Urban Studies." *Urban Studies* 42:1969–1990.

Moulaert, F., A. Rodriguez, and E. Swyngedouw, eds. 2003. *The Globalized City: Economic Restructuring and Social Polarization in European Cities*. Oxford: Oxford University Press.

Mudde, C. 2004. "The Populist Zeitgeist." *Government and Opposition* 39:542–563.

Nadasdy, P. 2007. "Adaptive Co-Management and the Gospel of Resilience." In *Adaptive Co-Management: Collaboration, Learning and Multi-Level Governance*, ed. D. Armitage, F. Berkes, and N. Doubleday, 208–227. Vancouver: University of British Columbia Press.

Nancy, J.-L. 1991. *The Inoperative Community*. Minneapolis: University of Minnesota Press.

Nancy, J. L. 1992. "La Comparution/The Compearance: From the Existence of 'Communism' to the Community of 'Existence.'" *Political Theory* 20:371–398.

Negri, A. 1990. "Postscript." In *Communists Like Us: New Spaces of Liberty, New Lines of Alliance*, ed. F. Guattari and A. Negri, 149–173. New York: Semiotext(e).

Nelson, S. H. 2015. "Beyond the Limits to Growth: Ecology and the Neoliberal Counterrevolution." *Antipode* 47:461–480.

New, M., D. Liverman, J. Schroder, and K. L. Anderson. 2011. "Four Degrees and Beyond: The Potential for a Global Temperature Increase of Four Degrees and Its Implications." *Philosophical Transactions of the Royal Society A: Mathematical Physical and Engineering Sciences* 369:6–19.

Newman, J. 2001. *Modernising Governance: New Labour, Policy and Society*. London: Sage.

Neyrat, F. 2008. *Biopolitique des Catastrophes*. Paris: Les Prairies Ordinaires.

Neyrat, F. 2010. "The Birth of Immunopolitics." *Parrhesia* 10:31–38.

Neyrat, F. 2014. "Critique du Géo-constructivisme Anthropocène & géo-ingénierie in Multitudes." *Multitudes* 56. http://www.multitudes.net/critique-du-geo-constructivisme-anthropocene-geo-ingenierie (accessed January 7, 2016).

Neyrat, F. 2016. *La Part Inconstructible de la Terre*. Paris: Editions du Seuil.

Neyrat, F., D. Johnson, and E. Johnson. 2014. "The Political Unconscious of the Anthropocene: A Conversation with Frédéric Neyrat." *Society and Space Open Site*. http://societyandspace.org/2014/03/20/on-8/ (accessed January 8, 2018).

Njeru, J. 2006. "The Urban Political Ecology of Plastic Bag Waste Problem in Nairobi, Kenya." *Geoforum* 37:1046–1058.

Novy, A., and B. Leubolt. 2005. "Participatory Budgeting in Porto Alegre: Social Innovation and the Dialectical Relationship of State and Civil Society." *Urban Studies* 42:2023–2036.

O'Malley, P. 1992. "Risk, Power and Crime Prevention." *Economy and Society* 21:252–275.

Oosterlynck, S., and E. Swyngedouw. 2010. "Noise Reduction: The Postpolitical Quandary of Night Flights at Brussels Airport." *Environment & Planning A* 42:1577–1594.

Paquet, G. 2001. "La Gouvernance en tant que Manière de Voir." In *La Démocratie à l'Epreuve de la Gouvernance*, ed. L. Cardinal and C. Andrew, 9–41. Ottawa: Presses de l'Université d'Ottawa.

Paddison, R. 2009. "Some Reflections on the Limitations to Public Participation in the Post-Political City." *L'Espace Politique.* http://espacepolitique.revues.org/index1393 .html (accessed July 10, 2010).

Pagden, A. 1998. "The Genesis of 'Governance' and Enlightenment Conceptions of the Cosmopolitan World Order." *International Social Science Journal* 50:7–15.

Peck, J. 2010. *Constructions of Neoliberal Reason.* Oxford: Oxford University Press.

Peck, J., N. Theodore, and N. Brenner. 2009. "Postneoliberalism and Its Malcontents." *Antipode* 41:94–116.

Pellow, D. 2007. *Resisting Global Toxics: Transnational Movements for Environmental Justice.* Cambridge, MA: MIT Press.

Pierre, J. 2000a. *Debating Governance: Authority, Steering and Democracy.* Oxford: University Press.

Pierre, J. 2000b. *Governance, Politics and the State.* Basingstoke: Macmillan.

Poulantzas, N. 1980a. "Research Note on the State and Society." *International Social Science Journal* 32:600–608.

Poulantzas, N. 1980b. *State, Power, Socialism.* London: Verso.

Prigogine, I., and I. Stengers. 1984. *Order out of Chaos: Man's New Dialogue with Nature.* Boulder, CO: New Science Library.

Protevi, J. 2013. *Life, War, Earth: Deleuze and the Sciences.* Minneapolis: University of Minnesota Press.

Pupovac, O., and S. Karamani. 2006. "On the Margins of Europe: An Interview with Rastko Močnik." *PRELOM—Journal for Images and Politics* 8:39–57.

Purcell, M. 2008. *Recapturing Democracy: Neoliberalization and the Struggle for Alternative Urban Futures.* New York: Routledge.

Raco, M. 2000. "Assessing Community Participation in Local Economic Development: Lessons for the New Urban Policy." *Political Geography* 19:573–599.

Raco, M. 2012. "A Growth Agenda without Growth: English Spatial Policy, Sustainable Communities and the Death of the Neo-liberal Project?" *GeoJournal* 77:153–165.

Rakodi, C. 2003. "Politics and Performance: The Implication of Emerging Governance Arrangements for Urban Management Approaches and Information Systems." *Habitat International* 27:523–547.

Rancière, J. 1989. *The Nights of Labor: The Workers' Dream in Nineteenth-Century France*. Philadelphia: Temple University Press.

Rancière, J. 1995a. *On the Shores of Politics*. London: Verso.

Rancière, J. 1995b. "Politics, Identification, and Subjectivization." In *The Identity in Question*, ed. J. Rajchman, 63–70. New York: Routledge.

Rancière, J. 1996. "Post-Democracy, Politics and Philosophy: An Interview with Jacques Rancière." *Angelaki* 1:171–178.

Rancière, J. 1998. *Disagreement*. Minneapolis: University of Minnesota Press.

Rancière, J. 2000a. "Dissenting Words. A Conversation with Jacques Rancière (with Davide Panagia)." *Diacritics* 30 (2):113–126.

Rancière, J. 2000b. *Le Partage du Sensible: Esthétique et Politique*. Paris: La Fabrique.

Rancière, J. 2001. "Ten Theses on Politics." *Theory & Event* 5 (3). https://muse.jhu.edu/article/32639 (accessed January 8, 2018).

Rancière, J. 2003a. "Comment and Responses." *Theory & Event* 6 (4). https://muse.jhu.edu/article/44787 (accessed February 6, 2018).

Rancière, J. 2003b. "Politics and Aesthetics: An Interview." *Angelaki* 8:194–211.

Rancière, J. 2004. "Introducing Disagreement." *Angelaki* 9:3–9.

Rancière, J. 2005. *Chroniques des Temps Consensuels*. Paris: Seuil.

Rancière, J. 2006a. *Hatred of Democracy*. London: Verso.

Rancière, J. 2006b. *The Politics of Aesthetics*. London: Continuum.

Rhodes, R. 1999. *Understanding Governance: Networks, Governance, Reflexivity and Accountability*. Buckingham: Open University Press.

Ricoeur, P. 1965. "The Political Paradox." In *History and Truth*, ed. P. Ricoeur, 247–270. Evanston: Northwestern University Press.

Robbins, P. 2007. *Lawn People: How Grasses, Weeds, and Chemicals Make Us Who We Are*. Philadelphia: Temple University Press.

Robson, M. 2005. "Introduction: Hearing Voices." *Paragraph* 28:1–12.

Rorty, R. 2004. "Post-Democracy." *London Review of Books* 26:10–11.

Rosanvallon, P. 2006. *Democracy Past and Future*. New York: Columbia University Press.

Rosanvallon, P. 2008. *Counter-Democracy*. Cambridge: Cambridge University Press.

Rose, N., and P. Miller. 1992. "Political Power beyond the State: Problematics of Government." *British Journal of Sociology* 43:173–205.

Royal Society, The. 2009. *Geoengineering the Climate: Science, Governance and Uncertainty*. London: The Royal Society.

Russo, A. 2006. "How to Translate Cultural Revolution." *Inter-Asia Cultural Studies* 7:673–682.

Sandercock, L. 1998. *Towards Cosmopolis*. Chicester, England; New York: J. Wiley and Sons.

Scherpe, K. R. 1987. "Dramatization and De-Dramatization of 'The End': The Apocalyptic Consciousness of Modernity and Post-Modernity." *Cultural Critique* 5:95–129.

Schlosberg, D. 2007. *Defining Environmental Justice: Theories, Movements and Nature*. Oxford: Oxford University Press.

Schmitt, C. 1996. *The Concept of the Political*. Chicago: University of Chicago Press.

Schmitter, P. 2000. "Governance." Paper presented at the Conference on Democratic and Participatory Governance: From Citizens to "Holders." European University Institute, Florence, 14 September.

Schmitter, P. 2002. "Participation in Governance Arrangements: Is There Any Reason to Expect It Will Achieve Sustainable and Innovative Policies in a Multi-Level Context." In *Participatory Governance: Political and Societal Implications*, ed. J. Grote and B. Gbipki, 51–69. Opladen: Leske and Budrich.

Sennett, R. 2007. *The Culture of the New Capitalism*. New Haven: Yale University Press.

Shellenberger, M., and T. Nordhaus. 2007. *Break Through: From the Death of Environmentalism to the Politics of Possibility*. Boston: Houghton Mifflin Co.

Showstack Sassoon, A. 1987. *Gramsci's Politics*. London: Hutchinson.

Simon, R. 1991. *Gramsci's Political Thought*. London: Lawrence & Wishart.

Skrimshire, S. 2010. *Future Ethics: Climate Change and Apocalyptic Imagination*. London: Continuum.

Sloterdijk, P. 2005. "Damned to Expertocracy." http://www.signandsight.com/features/238.html (accessed July 10, 2010).

Smith, N. 1984. *Uneven Development: Nature, Capital and the Production of Space*. Oxford: Blackwell.

Smith, N. 2008a. "Afterword." In *Uneven Development*, 3rd ed., ed. N. Smith, 239–266. London: University of Georgia Press.

Smith, N. 2008b. *Uneven Development: Nature, Capital and the Production of Space.* 3rd ed. with a new afterword. Athens: University of Georgia Press.

Springer, S. 2010. "Public Space as Emancipation: Meditations on Anarchism, Radical Democracy, Neoliberalism and Violence." *Antipode* 43:525–562.

Srnicek, N. 2008. "What Is to Be Done? Alain Badiou and the Pre-Evental." *Symposium* 12 (2): 110–126.

Staeheli, L. A. 2008. "Political Geography: Difference, Recognition, and the Contested Terrains of Political Claims-Making." *Progress in Human Geography* 32:561–570.

Staeheli, L. A., and D. Mitchell. 2008. *The People's Property: Power, Politics and the Public.* New York: Routledge.

Stavrakakis, Y. 1997. "Green Fantasy and the Real of Nature: Elements of a Lacanian Critique of Green Ideological Discourse." *Psychoanalysis, Culture & Society* 2:123–132.

Stavrakakis, Y. 2000. "On the Emergence of Green Ideology: The Dislocation Factor in Green Politics." In *Theory and Political Analysis: Identities, Hegemonies and Social Change*, ed. D. Howarth, A. J. Norval, and Y. Stavrakakis, 100–118. Manchester: Manchester University Press.

Stavrakakis, Y. 2006. *The Lacanian Left: Psychoanalysis, Theory, Politics.* Edinburgh: Edinburgh University Press.

Stoker, G. 1998. "Public-Private Partnerships in Urban Governance." In *Partnerships in Urban Governance: European and American Experience*, ed. J. Pierre, 34–51. Basingstoke: Macmillan.

Swyngedouw, E. 1996a. "The City as a Hybrid: On Nature, Society and Cyborg Urbanisation." *Capitalism, Nature, Socialism* 7:65–80.

Swyngedouw, E. 1996b. "Reconstructing Citizenship, the Re-Scaling of the State and the New Authoritarianism: Closing the Belgian Mines." *Urban Studies* 33:1499–1521.

Swyngedouw, E. 1997. "Neither Global nor Local: 'Glocalization' and the Politics of Scale." In *Spaces of Globalization: Reasserting the Power of the Local*, ed. K. Cox, 137–166. New York: Guilford Press.

Swyngedouw, E. 2000. "Authoritarian Governance, Power and the Politics of Rescaling." *Environment and Planning. D, Society & Space* 18:63–76.

Swyngedouw, E. 2004a. "Globalisation or 'Glocalisation'? Networks, Territories and Rescaling." *Cambridge Review of International Affairs* 17:25–48.

Swyngedouw, E. 2004b. *Social Power and the Urbanisation of Water: Flows of Power.* Oxford: Oxford University Press.

Swyngedouw, E. 2006. "Circulations and Metabolisms: (Hybrid) Natures and (Cyborg) Cities." *Science as Culture* 15:105–121.

Swyngedouw, E. 2007a. "Impossible/Undesirable Sustainability and the Post-Political Condition." In *The Sustainable Development Paradox,* ed. J. R. Krueger and D. Gibbs, 13–40. New York: Guilford Press.

Swyngedouw, E. 2007b. "The Post-Political City." In *Urban Politics Now: Re-Imagining Democracy in the Neoliberal City,* ed. BAVO, 58–76. Rotterdam: Netherlands Architecture Institute.

Swyngedouw, E. 2009a. "The Antinomies of the Postpolitical City: In Search of a Democratic Politics of Environmental Production." *International Journal of Urban and Regional Research* 33:601–620.

Swyngedouw, E. 2009b. "The Zero-Ground of Politics: Musings on the Post-Political City." *New Geographies* 1:52–61.

Swyngedouw, E. 2010a. "Apocalypse Forever? Post-Political Populism and the Spectre of Climate Change." *Theory, Culture & Society* 27:213–232.

Swyngedouw, E. 2010b. "The Communist Hypothesis and Revolutionary Capitalisms: Exploring the Idea of Communist Geographies for the Twenty-First Century." *Antipode* 41:298–319.

Swyngedouw, E. 2011. "Interrogating Post-Democratization: Reclaiming Egalitarian Political Spaces." *Political Geography* 30:270–280.

Swyngedouw, E. 2013a. "Apocalypse Now! Fear and Doomsday Pleasures." *Capitalism, Nature, Socialism* 24:9–18.

Swyngedouw, E. 2013b. "The Non-Political Politics of Climate Change." *ACME: An International E-Journal for Critical Geographies* 12:1–8.

Swyngedouw, E. 2014. "Anthropocenic Politicization: From the Politics of the Environment to Politicizing Environments." In *Green Utopianism: Politics, Practices and Perspectives,* ed. J. Hedrén and K. Bradley, 23–37. London; New York: Routledge.

Swyngedouw, E. 2015. "Depoliticized Environments and the Promises of the Anthropocene." In *International Handbook of Political Ecology,* ed. R. Bryant, 131–145. London: E. Elgar.

Swyngedouw, E. 2017a. "Insurgent Urbanity and the Political City." In *Ethics of the Urban: The City and the Spaces of the Political,* ed. M. Moshsen, 46–74. Zurich: Lars Müller Publishers/Harvard University Graduate School of Design.

Swyngedouw, E. 2017b. "More-than-Human Constellations as Immunological Bio-Political Fantasy in the Urbicene." *New Geographies* 9: 18–23.

Swyngedouw, E., and H. Ernston. 2018. "Interrupting the Anthropo-Obscene: Immuno-Biopolitics and the Re-Invention of the Political in the Anthropocene." Forthcoming in *Theory, Culture & Society*.

Swyngedouw, E., F. Moulaert, and A. Rodriguez. 2002. "Neoliberal Urbanization in Europe: Large Scale Urban Development Projects and the New Urban Policy." *Antipode* 34:542–577.

Swyngedouw, E., and E. Swyngedouw. 2009. "The Congolese Diaspora in Brussels and Hybrid Identity Formation: Multi-Scalar Identity and Cosmopolitan Citizenship." *Urban Research and Practice* 2:68–90.

Thomson, A. J. P. 2003. "Re-Placing the Opposition: Rancière and Derrida." Paper presented at the Fidelity to the Disagreement: Jacques Rancière and the Political conference, Goldsmith's College, University of London, September 16–17.

Valentine, J. 2005. "Rancière and Contemporary Political Problems." *Paragraph* 28:46–60.

Van Puymbroeck, N., and S. Oosterlynck. 2014. "Opening Up the Post-Political Condition: Multiculturalism and the Matrix of Depoliticisation." In *The Post-Political and Its Discontents: Spaces of Depoliticisation, Spectres of Radical Politics*, ed. J. Wilson and E. Swyngedouw, 86–108. Edinburgh: Edinburgh University Press.

Varoufakis, Y. 2017. *Adults in the Room: My Battle with Europe's Deep Establishment*. London: Bodley Head.

Velicu, I., and M. Kaika. 2015. "Undoing Environmental Justice: Re-Imagining Equality in the Rosia Montana Anti-Mining Movement." *Geoforum*. doi:10.1016/j.geoforum.2015.10.012.

Verdeil, E. 2014. "The Energy of Revolts in Arab Cities: The Case of Jordan and Tunisia." *Built Environment* 40:128–139.

Vergopoulis, C. 2001. "Globalisation and Post-Democracy." In *Welfare State and Democracy in Crisis: Reforming the European Model*, ed. T. Pelagidis, J. Milios, and L. T. Katseli, 37–50. London: Ashgate.

Virilio, P. 1986. *Speed and Politics: An Essay on Dromology*. New York: Semiotext(e).

Virno, P. 2004. *A Grammar of the Multitude*. New York: Semiotext(e).

Walker, G. 2009. "Beyond Distribution and Proximity: Exploring the Multiple Spatialities of Environmental Justice." *Antipode* 41:614–636.

Walker, G. 2012. *Environmental Justice: Concepts, Evidence and Politics*. London: Routledge.

Walker, J., and M. Cooper. 2010. "Genealogies of Resilience: From Systems Ecology to the Political Economy of Crisis Adaptation." *Security Dialogue* 42:143–160.

Ward, K. 2007. "'Creating a Personality for Downtown': Business Improvement Districts in Milwaukee." *Urban Geography* 28:781–808.

White, D. 2008. *Bookchin: A Critical Appraisal*. London: Pluto Press.

Whitehead, M. 2003. "'In the Shadow of Hierarchy': Meta-Governance, Policy Reform and Urban Regeneration in the West Midlands." *Area* 35:6–14.

Williams, E. C. 2011. *Combined and Uneven Apocalypse*. Washington, DC: Zero Books.

Williams, R. 1980. *Problems in Materialism and Culture*. London: Verso.

Williams, R. 1988. *Keywords*. London: Fontana.

Wilson, J., and E. Swyngedouw. 2015. *The Post-Political and Its Discontents: Spaces of Depoliticization, Spectres of Radical Politics*. Edingburgh: Edinburgh University Press.

Wilson, J., and E. Swyngedouw. 2015. "Seeds of Dystopia: Post-Politics and the Return of the Political." In *The Post-Political and Its Discontents: Spaces of Depoliticization, Specters of Radical Politics*, ed. J. Wilson and E. Swyngedouw, 1–22. Edinburgh: Edinburgh University Press.

Žižek, S. 1989. *The Sublime Object of Ideology*. London: Verso.

Žižek, S. 1992. *Looking Awry: An Introduction to Jacques Lacan through Popular Culture*. Cambridge, MA: MIT Press.

Žižek, S. 1999. *The Ticklish Subject: The Absent Centre of Political Ontology*. London: Verso.

Žižek, S. 2000. *The Fragile Absolute*. London: Verso.

Žižek, S. 2002a. *Revolution at the Gates: Žižek on Lenin, the 1917 Writings*. London: Verso.

Žižek, S. 2002b. *Welcome to the Desert of the Real*. London: Verso.

Žižek, S. 2006a. "Against the Populist Temptation." *Critical Inquiry* 32 (32): 551–574.

Žižek, S. 2006b. "The Lesson of Rancière." In *The Politics of Aesthetics*, ed. J. Rancière, 69–79. London: Continuum.

Žižek, S. 2006c. *The Parallax View*. Cambridge, MA: MIT Press.

Žižek, S. 2008a. "Censorship Today: Violence, or Ecology as a New Opium for the Masses." *Fordiletante*. http://fordiletante.wordpress.com/2008/05/07/censorship-today-violence-or-ecology-as-a-new-opium-forthe-masses (accessed August 5, 2012).

Žižek, S. 2008b. *In Defense of Lost Causes*. London: Verso.

Žižek, S. 2008c. *Violence.* London: Profile Books.

Žižek, S. 2013a. *Demanding the Impossible.* Ed. Yong-june Park. Cambridge: Polity Press.

Žižek, S. 2013b. *The Idea of Communism 2: The New York Conference.* London: Verso.

Žižek, S. 2017. *The Courage of Hopelessness.* London: Allen Lane.

Index

Power (cont.)
  technologies of, 13
  of those who do not count, 131
Proletarian, the, 168
Proletariat movement, 44, 51, 105–106,
    133, 142, 151
Property rights, 153, 159–161, 163
Public space, 26–27, 54

Raco, Mike, 31
Rancière, Jacques, 23, 28–29, 31, 41,
    46–52, 54–55, 57, 99, 106, 107, 119,
    130, 136
Real, the
  of all things, 71
  of capital, 164
  of class antagonism, 44
  of ecological disintegration, 86–88
  of natures, 79, 84
  of the political, 58, 144
  the political and, 43–44, 144
  of the present, 62
Real Democracy Now!, 130, 133
Refugees, 164–167
Representation in post-democratization
    process, 16–17
Resistance
  hysterical act of, 58, 144
  rituals and acts of, 57–58
Ricouer, Paul, 40
Right holders, 9
Robbins, Paul, 117
Robespierre, Maximilien, 56
Romanticism, 78
Rorty, Richard, 30
Rosanvallon, Pierre, 26–27, 41

Saramago, José, xv
Sarkozy, Nicolas, 58
Saurí, David, 117
Scherpe, Klaus, 99
Schmitter, Philippe, 7, 8

Scum, 58, 131
Sensible, the, 51–52, 130
Serres, Michel, 73
Shareholders, 9
Sloterdijk, Peter, 158
Smith, Neil, 103
Socialism, xx, 61, 147, 153–154,
    157–158
*Social Power and the Urbanization of Na-
    ture* (Swyngedouw), 117
Society
  articulating state, market and civil,
    9–14
  "good," 32
Solidarnosc, 56–57
Space
  democratizing, 136–138
  dissensual, the political as, 52–54
  egalitarian, producing, 54–62, 145
  egalitarian, staging, 46–54, 142
  repoliticizing, 59
Space holders, 9
Stakeholders, 9, 16
State
  capitalist class and the, 27
  death of the, 20
  in government/governance configura-
    tion, shift to, 20–22
  legitimacy of the, 10
  privatization of the, 157–158
State/market/civil society relationship,
    9–14, 27
State of emergency, a permanent, 34–35
Status holders, 9
Stavrakakis, Yannis, 42
Stiglitz, Joseph, 28
Sustainability, 67, 81–88, 109
Swyngedouw, Erik, 117

Taksim Square revolt, Istanbul, 37, 122,
    129, 134
Taylorism, 162

www.ingramcontent.com/pod-product-compliance
Lightning Source LLC
Chambersburg PA
CBHW030837300326
41935CB00037B/498